PROFILES IN MATHEMATICS

Pierre de Fermat

Profiles in Mathematics:

Pierre de Fermat

Chad Boutin

MORGAN REYNOLDS

PUBLISHING

Greensboro, North Carolina

Profiles in Mathematics:

Alan Turing

Rene' Descartes

Carl Friedrich Gauss

Sophie Germain

Pierre de Fermat

Ancient Mathematicians

Women Mathematicians

PROFILES IN MATHEMATICS
PIERRE DE FERMAT

Copyright © 2009 By Chad Boutin

Library of Congress Cataloging-in-Publication Data

Boutin, Chad.
    Pierre de Fermat / by Chad Boutin.
        p. cm. -- (Profiles in mathematics)
    Includes bibliographical references and index.
    ISBN-13: 978-1-59935-061-5
    ISBN-10: 1-59935-061-0
    1. Fermat, Pierre de, 1601-1665. 2. Mathematicians--France--Biography.
I. Title.
    QA29.F45B68 2009
    510.92--dc22
    [B]

                                    2008015765

Printed in the United States of America
First Edition

Contents

Introduction

Mathematics gives us a powerful way to analyze and try to understand many of the things we observe around us, from the spread of epidemics and the orbit of planets, to grade point averages and the distance between cities. Mathematics also has been used to search for spiritual truth, as well as the more abstract question of what is knowledge itself.

Perhaps the most intriguing question about mathematics is where does it come from? Is it discovered, or is it invented? Does nature order the world by mathematical principles, and the mathematician's job is to uncover this underlying system? Or is mathematics created by mathematicians as developing cultures and technologies require it? This unanswerable question has intrigued mathematicians and scientists for thousands of years and is at the heart of this new series of biographies.

The development of mathematical knowledge has progressed, in fits and starts, for thousands of years. People from various areas and cultures have discovered new mathematical concepts and devised complex systems of algorithms and equations that have both practical and philosophical impact.

To learn more of the history of mathematics is to encounter some of the greatest minds in human history. Regardless of whether they were discoverers or inventors, these fascinating lives are filled with countless memorable stories—stories filled with the same tragedy, triumph, and persistence of genius as that of the world's great writers, artists, and musicians.

Knowledge of Pythagoras, René Descartes, Carl Friedrich Gauss, Sophie Germain, Alan Turing, and others will help to lift mathematics off the page and out of the calculator, and into the minds and imaginations of readers. As mathematics becomes more and more ingrained in our day-to-day lives, awakening students to its history becomes especially important.

Sharon F. Doorasamy
Editor in chief

Editorial Consultant

In his youth, Curt Gabrielson was inspired by reading the biographies of dozens of great mathematicians and scientists. "I was driven to learn math when I was young, because math is the language of physical science," says Curt, who named his dog Archimedes. "I now know also that it stands alone, beautiful and mysterious." He learned the more practical side of mathematics growing up on his family's hog farm in Missouri, designing and building structures, fixing electrical systems and machines, and planning for the yearly crops.

After earning a BS in physics from MIT and working at the San Francisco Exploratorium for several years, Curt spent two years in China teaching English, science, and math, and two years in Timor-Leste, one of the world's newest democracies, helping to create the first physics department at the country's National University, as well as a national teacher-training program. In 1997, he spearheaded the Watsonville Science Workshop in northern California, which has earned him recognition from the U.S. Congress, the California State Assembly, and the national Association of Mexican American educators. Mathematics instruction at the Workshop includes games, puzzles, geometric construction, and abacuses.

Curt Gabrielson is the author of several journal articles, as well as the book *Stomp Rockets, Catapults, and Kaleidoscopes: 30+ Amazing Science Projects You Can Build for Less than $1.*

Pierre de Fermat
(Courtesy of Visual Arts Library (London)/Alamy)

one
Early Years

Pierre Fermat (pronounced *fur-mah*) was born around August 20, 1601, in the small town of Beaumont-de-Lomagne, about twenty-five miles northwest of Toulouse, a city in southwest France on the banks of the Garonne River. Many details of his life are unknown or uncertain, so his story, the story of one of the world's greatest mathematicians, does not come as an unbroken curve, as a mathematician might describe it, but as a series of discontinuous points.

Pierre Fermat's father, Dominique, was a prosperous leather merchant whose wife, Claire de Long, came from a family prominent in the legal business. They had four children—a son and two daughters in addition to Pierre—and the family grew steadily in wealth and prominence during Pierre's childhood. Though they were quite prosperous by the time Pierre was a young man, the Fermats were not members of the nobility. They sought to increase their social status not only by amassing wealth, but also by providing their children with an education that could open doors to a better life.

Nearby Toulouse was famous for its university, but its position as a commercial center had declined in recent years. What kept its economy going was the volume of legal work that took place in town, both in the *parlement* (a regional court) and in the many lower courts there. A career in law would have been a natural choice for a talented boy whose family already had connections to the profession, and at some point his parents destined young Pierre for legal training. Along with children from other wealthy families, he was probably sent to the nearby Franciscan monastery at Grandselve for his first years of school.

Catholic friars were dedicated to an austere life of religious devotion, and in exchange for the support of the community they offered to teach the local children how to read and write, skills that the friars had kept alive during the Middle Ages. For centuries, the only way to preserve books was to copy them by hand, and were it not for the work of monks, many writings by ancient authors would have been lost to civilization.

These ancient authors were held in the highest respect by virtually all educated people, who felt that the pinnacle of civilization had been reached during Roman and Greek times. At times this respect bordered on awe, and when learning started to revive during the Renaissance, artists and scholars began to spend much of their time admiring the forms of ancient works, all the while trying to recover the techniques the ancients had used to create their statues and architecture. The lesson a student heard again and again was: *the ancients have learned everything valuable that can be learned, and we cannot hope to improve upon their work. The best we can do is absorb what they have left us and try to restore some of what was lost.*

Pierre and his classmates spent most of their time mastering languages, literature, history, and the rhetoric of ancient speeches. Cicero, the renowned Roman orator, was held in the highest regard, and Pierre read many of Cicero's speeches as part of his education. He read these with extra care, for Cicero's style would be useful to any lawyer.

From his linguistic studies, Pierre understood five languages: in addition to his native French, he knew Spanish, Italian, Greek, and Latin. By the time he was a young man, he would have had a thorough grounding in many ancient writers besides Cicero, including Greek philosophers such as Plato and Pythagoras, having read them all in the original language.

In Pierre's time, Latin was especially important because it was the language of communication in the Catholic Church as well as the language by which educated Europeans in general communicated. Serious books had little chance of finding an audience unless they were written in Latin. And with the invention of the printing press several decades before Pierre's birth, serious books were being distributed and read in increasing numbers.

After finishing his primary education, Pierre Fermat needed to go to university for a law degree. Evidence suggests he attended the University of Toulouse for a time. For a future lawyer, Toulouse was an excellent place to live. The university was one of the oldest in Europe, and it attracted students from as far away as Poland. Often they came to study Roman

Cicero *(Courtesy of Visual Arts Library (London)/Alamy)*

law, as the University of Toulouse was one of the few places where that legal system was taught.

Toulouse was even home to Europe's oldest literary society, and composing verse was an activity for a gentleman of the time. Typical of a bright young person with a classical education, Pierre tried his hand at composition, and among his papers survive a few of his poems—written in Latin, French, and Spanish.

Ironically, one subject that he probably encountered very little of at the university was

Euclid

mathematics. During Pierre's early life, math was not a subject one could study in depth at any European university. A student read the first few books of Euclid's *Elements*, which for nearly 2,000 years had remained the primary textbook on geometry. But then the student would be done with the subject.

Euclid's books were quite short. An ancient book was actually a roll of papyrus sheets fifteen to twenty feet long that contained about 15,000 words on average. Only a paltry few ancient mathematics texts survived into Pierre Fermat's day—offering an incomplete record of what the Greeks had accomplished, and precious little on how they had accomplished it. Careful reading and long meditation would be needed to rebuild mathematical knowledge from such scanty resources. When Pierre was born, this work to restore the mathematical past had only just begun.

Nonetheless, Pierre was surely impressed even with the little ancient math he had seen. Euclid systematically introduced him to the construction and manipulation of lines, geometric shapes, and curves, as well as the solutions to some problems relating to them. Had Fermat read further, the last two books of the *Elements* would have exposed him to the work of Pythagoras, who first introduced to the world the idea of the proof, a demonstration of a statement's truth or falsehood based on logical reasoning. Euclid employed proofs to demonstrate the ideas in the *Elements*, and both he and Pythagoras viewed the search for mathematical truth as an end in itself. Even if no practical applications resulted, contemplation of the world of numbers revealed a sublime beauty that justified a lifetime of effort.

By Pierre Fermat's time, virtually no one considered searching for the hidden relationships among numbers their life's work. There were several reasons why, in addition to the comparative lack of books on math that had survived into the early seventeenth century. Even among those few who appreciated its value, there was no agreement as to what precisely mathematics was, what sort of problems should be explored, or which methods should be used for doing so. Some believed the study of geometry to be valuable for the design of buildings, while astrologers wanted tools for making predictions about the future based on planetary motion. Others found math to be an exercise in pure contemplation. From such divergent points of view emerged no compelling reason to train students in any sort of math.

In any case, Pierre would have known that mathematics was not something one did for a living. It was at best a diversion, something one could do to take the mind off the business of daily existence.

Complicating things further, Pierre lived in a period of history marked by trouble, especially for people with new ideas. He grew up in a time when France was still ruled by a king. The royal family, headed by King Louis XIII during Pierre's early life, claimed absolute power under the authority of God and the Catholic Church. The Catholic Church also exercised a strong influence in France and across the rest of Western Europe. This influence, which had persisted for many centuries, did not go unchallenged, and Toulouse had been affected by the religious wars that had swept the region in the preceding century. Though Toulouse was home to a large, respected university, the Dominican priests influenced the faculty's activities. The renewal of learning and free inquiry that the Renaissance had brought to Europe grew with difficulty in the provincial town.

Toulouse was still important to the monarchy, which had obtained the support of the city only after a prolonged struggle. Religious wars in Languedoc, the province of which Toulouse was the capital, had ended a mere five years before

A nineteenth-century view of Toulouse, France *(Library of Congress)*

Pierre's birth, in 1596—the year when the city finally recognized Henry IV as king.

Because the Fermat family was prominent in both the mercantile and legal communities, it is probable that Pierre was well aware of the religious and political tensions that existed during his formative years. In 1619, for instance, a well-educated philosopher named Lucilio Vanini, once a popular tutor for children of the wealthy, was charged with blasphemy and atheism. He was burned at the stake in February. Pierre was not yet eighteen.

The following years of Pierre Fermat's life are mostly lost to history. After finishing his education at the monastery, it is unknown whether or not he attended the University of

Toulouse, and if he did attend, he left without a diploma. A decade or more would pass before he would finally obtain the law degree he would need to begin work. What took such a talented young man away from school for so long can only be guessed at. It was not lack of money; his family had plenty. It was probably not lack of direction, either; his family would have encouraged him at every turn to finish his legal education.

One possibility is that Pierre Fermat knew on some level that he needed more than just a respectable position in society. His curiosity about mathematics may have driven him to find out more than he could ever have learned in school, and for that he would have to find someone who had mastered the subject well enough to teach him. The effort to find such a person carried Fermat on the longest journey he would ever take.

two

Meeting the Master

At some point in the 1620s, Pierre Fermat traveled a hundred miles down the Garonne River from Toulouse to the city of Bordeaux, where his family had friends who were also involved in legal work. Fermat does not seem to have pursued any formal legal studies while he was there, but he stayed for several years.

Though the purpose of this journey remains unknown to history, what is certain is that during his sojourn away from home—probably the farthest he would ever travel from Toulouse, and unquestionably for the longest time—he encountered the mind that would most influence his own work. This was François Viète (pronounced *vyeh*-tuh), a visionary mathematician in his own right who had died in 1603, more than twenty years previous. Some of Viète's former students lived in Bordeaux and had preserved their teacher's legacy of knowledge. Before long, the young Fermat

François Viète

Fermat lived in Bordeaux, France for several years in the 1620s. *(Library of Congress)*

made contact with them and began to absorb the deceased mathematician's work.

Viète had a few things in common with Fermat. He had also grown up in relatively comfortable circumstances and been trained in the law. His intelligence, hard work, and connections in high society had brought him a great deal of professional success, eventually leading to his appointment as a royal privy councilor.

Viète also loved exploring the world of numbers. Most of his life, however, his responsibilities forced him to pursue his passions as a hobby, if he had time to pursue them at all. Luckily, the king recognized his talents and asked him on occasion to solve problems no one else could handle, such as deciphering coded messages. Viète was so good at recognizing patterns that some accused him of employing sorcery to crack difficult codes. Living in Bordeaux like Viète and walking the same streets as the master mathematician must have been exciting for Fermat: he had both the leisure and the opportunity to delve deeply into the subject he loved best. Finally, he could get far more than a mere glimpse at ancient mathematical ideas that few around him understood or cared about. He could read the ideas of a modern thinker whose writings were, compared to those of the Greeks, still wet on the page. And because a few of Viète's students lived in Bordeaux, he even had like-minded people to talk to.

Toward the end of his life, Viète had spent several years in an attempt to forge effective tools that a mathematician could use for thinking and problem solving. Fermat pondered these tools until he fully understood them and could add them to his own mathematical toolbox. But Viète's efforts had gone only so far. As Fermat explored the gap between how much the ancients knew and how little of it the centuries had preserved, he must have realized how much more work there was to be done.

Viète, like many educated men of his own and Fermat's day, accepted that one of his most important tasks was to restore as much as he could of the work of the ancient masters. Euclid's great *Elements* had survived, but most of the ancients' work had been lost irretrievably. Sometimes all that remained of a Greek thinker's work was his name in someone else's book, along with a line or two mentioning that he had been able to solve some difficult type of problem. From such scant information, mathematicians sought to re-create what had been known to their predecessors fifteen hundred years earlier.

Of course, this was not as easy as translating from one language to another. To solve the old problems, entirely new techniques often had to be created from whatever mathematical tools a person could manage to find and then master. But which tools? And what approach? Since mathematicians of the period did not agree about what to study or why, the road was far from clear.

Another obstacle was that the few ancient math collections that had survived were typically filled with interesting insights into many individual problems, but no comprehensive method for solving all problems of a certain type. In other words, the ancients had provided many specific solutions, but they had rarely provided general solutions or rules. An ancient author who bucked this trend was Pappus, one of two Greek mathematicians whose surviving work became important to both Viète and Fermat.

Pappus lived in the city of Alexandria, in Egypt, in the fourth century CE., a time when few mathematicians were left to pass along what had been learned over the years. For the sake of the future, he decided to write a handbook called the *Mathematical Collection*. He hoped the work would help people grasp the ideas of other ancient mathematicians even if no living teacher could be found. Pappus also surveyed the important works of several other mathematicians, an effort that helped individuals like Viète reconstruct what these other ancients had known.

The *Mathematical Collection* was particularly useful to Viète because it contained a general description of the two main methods of analysis by which the Greeks solved problems. When Viète read through these methods in the 1580s, he noticed that Greek writings might help him in his own work. He substantially reworked them into three techniques of his own. These techniques were called:

• *Zetetics*: transforming a problem into an equation that links an unknown value with various known values

• *Exegetics*: transforming the equation found by *zetetics* so as to determine the unknown value

• *Poristics*: exploring the truth of a theorem by symbolic manipulation.

These techniques are today known as algebra, and Viète was the first person to successfully link algebra with Greek analysis.

Of the many different branches of mathematical study that existed in the seventeenth century, algebra was probably the least respected; nothing profound had yet been discovered by using it. Though its roots extend as far back as Babylonia, and Arab mathematicians had employed it for centuries during the medieval period (the word *algebra* itself comes from Arabic), it had survived primarily as a tool for assisting merchants with financial transactions.

Viète, however, sensed algebra's potential for solving many types of problems in geometry and arithmetic. He realized that this disrespected form of calculation might be a key to rediscovering the power of the Greeks to solve such problems—even if the Greeks themselves had used a completely different method.

Mathematical relationships in Viète's day were often expressed in lengthy verbal statements, sometimes stretching out for a paragraph, rather than with compact equations that required a single line. Viète perceived that algebra had the power to overcome these difficulties.

In perhaps his most significant contribution to the advance of mathematics, Viète hit upon the idea that letters could be used to represent quantities. He decided to use consonants for known values and vowels for unknowns. He also adopted the plus and minus symbols we use today. But he did not create a complete symbolic language; he still used a few words to express ideas such as exponents, multiplication, and equality. Whereas we would express the general form of a second-degree equation by the notation

$$Ax^2 + Bx = C$$

Viète would have expressed it as

$$\textbf{D } \textit{in}\textbf{ E } \textit{quadratum} + \textbf{F } \textit{in}\textbf{ E } \textit{aequetus}\textbf{ G.}$$

While not quite as easy to work with as the mathematical notation used today, this was a great improvement over spelling everything out. A few letters, numbers, and symbols in a single line could express the same concept far more succinctly, and in a form that mathematicians could easily manipulate. Viète had begun to create a powerful tool for solving general problems far beyond the needs of the mercantile world.

Fermat pored over the papers Viète had left behind, noticing how different the ideas were from those found in Euclid. He wondered if the new approach could help him understand what Euclid did, and how he did it. Fortunately, Fermat was not alone.

Among Viète's remaining followers in Bordeaux, some lent support to Fermat's efforts, and Fermat mentions them in early letters to a friend. He describes Etienne d'Espagnet, who was the son of a member of the Bordeaux *parlement* and ten years older than Fermat, as a close friend. D'Espagnet possessed some of Viète's papers and gave many of them over to Fermat by 1638. Jean Beaugrand was another older man who saw himself as Fermat's mentor. A few years later, he would play a role in introducing Fermat to the larger mathematical community, though the friendship would also cause Fermat some trouble. At the time, however, Beaugrand and Fermat developed a good working relationship. Beaugrand was well respected in Bordeaux for his own mathematical talents.

In any case, enough clues were left behind for Fermat to do a great deal of work alone. Viète showed that his symbolic language could handle many problems from both contemporary and ancient sources.

Despite his importance, in retrospect, few mathematicians of his day perceived Viète's significance. He had published a few papers in the 1590s that had reached Paris. But when at the age of sixty-three Viète died, none of the major thinkers of the period followed his lead. None, that is, until Fermat arrived in Bordeaux about a quarter century later. And Fermat was sufficiently impressed to immerse himself fully in Viète's world—a world in which problems had clear solutions and everything tended toward harmony and balance.

Nearly every one of Fermat's major achievements spring from his time in Bordeaux. By the time he left, most of his great innovative ideas were present in his mind, if only in embryonic form. From their inspiration he found direction for the mathematical roads he would travel for the rest of his life.

As he read through Viète's papers, Fermat was drawn to what Pappus had done with one particular ancient book: the *Plane Loci* of Apollonius, which contained 147 problems regarding straight and curved lines. (A *locus*, the singular form of *loci*, is simply a line, curve, or geometric shape or solid.) The sheer number of these problems had suggested even to the Greeks that a way to simplify or generalize this knowledge was needed, and Pappus managed to distill the original 147 problems down to sixteen general cases.

History has respected the value of Pappus's approach. To this day, mathematicians seek to do more than just solve individual problems; they seek methods for approaching entire classes of problems.

Eventually Fermat was able to distill the sixteen cases of Pappus down to a single comprehensive theorem about curves—discussed at length in his work *Introduction to Plane and Solid Loci*. Though this paper did not come to wider attention until several years later, Fermat finished much of the *Introduction* before leaving Bordeaux, beginning work in about 1628. By 1629, though he was still struggling to express some of his ideas, he had compiled a draft of his work, and he left a more or less finished version with his friends in Bordeaux when he departed for Orléans, France, to continue his schooling the following year.

As so often is the case with Fermat's work, his efforts went further than his original intention. The *Introduction* was more than just a restoration of the past; it was a new and modern approach to ancient problems. Fermat had taken his first step toward creating the math used today; he took another step when he came across the other ancient mathematician who had inspired Viète.

That mathematician was Diophantus, who along with Pappus ranks as one of the two great Greek mathematicians of late Alexandria. Diophantus is believed to have been born around 200 C.E., nearly a century before Pappus, though there is much uncertainty about the details of his life. Only fragments of his work survive, but these fragments contain some of the earliest known forms of algebraic computation. Viète essentially took Pappus's description of methods and joined them to procedures he encountered in Diophantus to create his own algebraic technique.

Only the first six books of Diophantus's thirteen-book *Arithmetica* were known in the seventeenth century (four more have since been discovered). But these were enough to keep Fermat coming back long after he had absorbed Viète's methods, for Diophantus opened Fermat's mind to another subject that proved an endless source of inspiration. The *Arithmetica* was a book on problem solving. Yet many of the problems it contained had nothing to do with the kinds of practical questions that algebra can be used to answer, such as how fast a falling object will be traveling at a certain time. Rather, the problems in the *Arithmetica* were often highly abstract:

- Find two numbers such that either one, when added to the square of their sum, gives a square.
- Find four numbers such that the product of any two of them, plus one unit, is a square.

> # DIOPHANTI
> ## ALEXANDRINI
> ### ARITHMETICORVM
> #### LIBRI SEX,
> *ET DE NVMERIS MVLTANGVLIS LIBER VNVS.*
>
> *Nunc primùm Græcè & Latinè editi, atque abfolutißimis Commentariis illuftrati.*
>
> AVCTORE CLAVDIO GASPARE BACHETO MEZIRIACO SEBVSIANO, V. C.
>
> ## LVTETIAE PARISIORVM,
> Sumptibus SEBASTIANI CRAMOISY, via Iacobæa, fub Ciconiis.
> ## M. DC. XXI.
> *CVM PRIVILEGIO REGIS.*

The title page of Diophantus' *Arithmetica*, which influenced Fermat's interest in abstract problem solving

• Find a right triangle such that the number of units in its area, when added to the number of units in the length of one smaller side, equals a given number.
• Find two numbers such that the cube of the greater plus the lesser is equal to the cube of the lesser plus the greater.

These are exercises in pure number manipulation, puzzles that for Fermat had their own intrinsic beauty and meaning. They exposed the sorts of relationships between numbers that had so fascinated Pythagoras and his followers. This was Fermat's introduction to what came to be called number theory. Before long, Fermat had his own copy of *Arithmetica* to study.

His head full of new ideas, Fermat was ready to return to the life that awaited him in Toulouse; the next few years show him full of resolve. He traveled to Orléans, where he at last obtained his law degree sometime before 1631. Fermat was already thirty, ten years older than many of his class-mates, but he returned to Toulouse and quickly settled down. In January 1631, he purchased his position in *parlement*, as all magistrates had to do, and the position entitled him to change his name. Pierre Fermat became Pierre de Fermat, nobleman and pillar of the community. Four months later, he married Louise de Long, his mother's fourth cousin. The outward course of the rest of his life was set.

three
Introduction to the Paris Circle

Toulouse was an enormous town, the second largest in all of France and the capital of its southernmost province. Its population had reached the astonishingly high figure of 40,000. The students at its eminent university numbered hundreds more, and the dozens of lawyers, priests, and governmental functionaries traveled its streets every day.

By the spring of 1636, Pierre de Fermat had spent five years as a family man and legal professional in Toulouse. He was well established in his career, and within two years he would be promoted to the Chamber of Investigations. His mathematical avocation, however, was still almost completely a private matter. Though he had been working steadily on several topics in math that had interested him since his time

in Bordeaux, he had spoken of his leisure pursuits c___
few acquaintances.

Of these acquaintances, none was more important than
Pierre de Carcavi. Carcavi was one of Fermat's fellow par-
liamentarians and, like Fermat, was keenly interested in the
mathematical and scientific research happening throughout the world. Fermat frequently spoke of his own work to
Carcavi, as his friend and colleague was one of the few people capable of understanding Fermat's work and interests. As
such, Fermat was disappointed when Carcavi announced that
he had bought a prestigious position in the *Grand Conseil*
(Great Council) of Paris, and would be moving to that city.
However, it was this move that helped spread Fermat's work
and reputation throughout the world.

In almost every way, Paris—one of Europe's biggest
cities—was the center of the world to a Frenchman. Far
larger even than Toulouse, it had long been the cultural
and economic center of France. It had become the country's scientific center as well; in the drawing rooms and
salons of the city, many of the most talented men in the
country gathered to discuss the latest discoveries that
practitioners of the new "natural philosophy" were turning
up. Carcavi was not a strong researcher or mathematician
in his own right, but he was deeply interested in scientific subjects, and he was a good judge of talent in others.
Within weeks of his arrival in Paris, he had met the man
most responsible for bringing the city's scientists together
as a community and uniting them with thinkers in other
countries. This man was Marin Mersenne, a priest of the
Minim, an order of Catholic priests who led particularly
Spartan lifestyles.

Marin Mersenne

Mersenne was famous across Europe as a gifted thinker who felt that truth—scientific truth included—was humanity's best defense against sin and error. He was quick to defend the Catholic Church from detractors, but he also believed science's search for new truth could reveal more of God's glory. In an age when it was dangerous to contradict Catholic teachings, Mersenne hoped religious doctrine and scientific discovery would gradually become less mutually antagonistic.

Carcavi told Father Mersenne of his talented friend Fermat who seemed to know so much about the properties of numbers, especially primes, not to mention ways of factoring large numbers. Mersenne was intrigued: one of the subjects that most interested him was music, and building tonally accurate instruments required knowledge of the relative lengths of vibrating strings or organ pipes, all of which were described with mathematical relationships.

Fermat had even discovered a new pair of so-called friendly numbers. These are two numbers such that the sum of all the divisors of the first number equals the second number, and vice versa. The Pythagoreans had known only that 220 and 284 were friendly, but Fermat had found two friendly numbers that were much larger.

Fermat's ability to manipulate a number's divisors (which he called aliquot parts) eventually made it into *Universal Harmony*, Mersenne's 1636 paper on music theory:

> Now, if I wished to speak of the men of high birth or quality who have thrived so well in this area of mathematics [music theory] that no one can teach them anything, I . . . would add Monsieur Fermat . . . to whom I owe the observation he made concerning the two numbers 17296 and 18416, the aliquot parts of each of which constitute the other, as do those of the two numbers 220 and 284; also concerning the number 672 [and the number 1344]. . . . He knows the infallible rules and the analysis for finding an infinity of others of this sort.

Intrigued by Carcavi's endorsement, Mersenne sent Fermat a letter inviting him to offer a few observations on free fall, a subject that Beaugrand, his old friend from Bordeaux, had confronted in a paper that was causing a bit of a stir in Paris.

Though Fermat went his whole life without ever once visiting the faraway capital, he was not unaware of life outside Toulouse, and he was astonished to receive a letter from the eminent Mersenne, who regularly corresponded with scientists from as far away as Syria. Fermat quickly went to work drafting a reply.

While Mersenne had merely requested Fermat's observations on the subject of free fall—a matter that had also concerned Galileo—Fermat replied with a good deal more. He not only addressed the subject of Beaugrand's paper, but also said that Galileo himself had been wrong in claiming that a free-falling object would follow a semicircular path to the center of the earth; Fermat enclosed an elegant proof that the path would instead be a spiral. He enclosed further work on spirals and told Mersenne that he had also restored the text of Apollonius's *Plane Loci*, mentioning offhandedly that he had found useful approaches to many other problems for which Viète's tools alone could not have been sufficient.

Tellingly, Fermat appended to this mountain of information a request for news of any developments in math that had occurred over the previous five or six years. Mersenne, impressed by the work Fermat had done, more or less in isolation from the other great mathematicians of the day, showed the letter to his Paris colleagues, and they recognized that into their midst had come a powerful mind that could delight and challenge them.

The letters Fermat sent to Paris over the next few months explained his ideas briefly and in piecemeal fashion, making it difficult for his correspondents to follow them. But his *Introduction to Plane and Solid Loci* served as a good overview of his Bordeaux accomplishments. He sent this paper to Paris probably in late 1636, a few months after the correspondence began.

By mid-1636 Fermat had begun corresponding regularly with Mersenne and two other members of the Parisian circle, Gilles Roberval and Etienne Pascal. The latter was a gifted scientist whose son Blaise became one of history's most influential thinkers.

Roberval became one of Fermat's longtime correspondents. He was particularly interested in what Fermat's methods could reveal about curves, such as their maximum and minimum values. Fermat's thinking had ranged widely on such subjects. His letter to Roberval on September 22, 1636, which references a paper he had left with d'Espagnet in Bordeaux, gives an idea of how Fermat's abilities had blossomed beyond describing conic curves and their extreme points:

> If M. d'Espagnet has set forth my method [of finding the maximum and minimum points of curves] to you only as I sent it to him at the time, you have not seen its most beautiful applications. For, with slight alterations, I make it serve
> 1. for finding propositions similar to those on the conoid [curve] which I sent you in my last letter;
> 2. for finding tangents to curved lines, on the subject of which I propose this problem to you: to find the tangent to a given point on the [conchoid] curve of Nicomedes;
> 3. for finding the centers of gravity of all sorts of figures, even figures different from the ordinary, like my conoid and infinitely many others, examples of which I will show you whenever you wish;
> 4. for numerical problems which deal with aliquot parts and which are all very difficult . . . .
> I have left out the main application of my method, which is for the determination of plane and solid loci.

Roberval quickly grew interested in Fermat's third point, regarding techniques for finding the centers of gravity of geometric figures. Another of Fermat's revolutionary ideas made this possible. The idea, which Fermat called *adequality*, became important in a mathematical controversy the following year.

A steady stream of Fermat's letters flowed into Paris on many of the topics that Fermat had been considering since his years in Bordeaux. However, despite his generosity with the knowledge he had acquired, Fermat was hesitant to reveal his methods for solving problems.

Fermat would typically post a few problems for the Paris group to solve (as he did with the second item of the letter to Roberval). He would mention that he was able to solve the problems by methods of his own, but he would not initially share his solutions. In a later letter, he would provide the solutions, but even then they were often abbreviated, requiring the reader to make substantial mental leaps in order to understand Fermat's thought process. These habits would come to irritate many of Fermat's correspondents throughout his career.

A taste of Fermat's approach, as well as the terse style of description that frustrated so many of his correspondents, can be found in the following example from the *Introduction*, which demonstrates the construction of a straight line (of the form in group 1 above):

> Let NZM be a straight line given in position, of which point N is given. Let NZ be equal to the unknown quantity A, and let the line ZI, erected at a given angle NZI, be equal to the other unknown quantity E. If D times A equals B times E, then point I will lie on a *straight line* given in position.
>
> That is, as B is to D so A is to E. Therefore, the ratio of A to E is given; and the angle at Z is given. Therefore, the triangle NIZ is given in species, and angle INZ [is given]. Also, point N is given, and the straight line NZ [is given] in position. Hence, NI will be given in position. The composition is easy.

This proof considers two points along the axis from which two line segments extend upward. In essence, it states that if the ratio of the two points' distances from the origin is the same as the ratio of the respective line segments' lengths, then the origin and the endpoints of the two segments lie on a straight line. The proof itself is complete by the end of

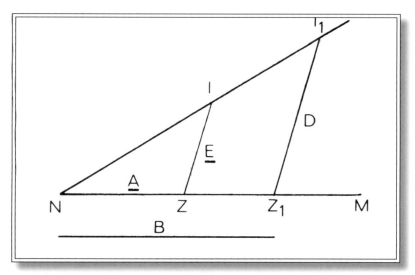

An example from the *Introduction*

the first paragraph quoted above, and the second paragraph is an analysis of it.

Though the concepts were relatively straightforward for Fermat's readers (especially when compared to the more complex proofs later in the *Introduction*), Fermat leaves out a few things that a seventeenth-century reader accustomed to the classical style of Greek proofs would have expected. For example, he aims to accomplish two goals here: first, to construct a line with an equation, and second, to show that point I lies on the line regardless of the values of A and E. In Greek math, these requirements would have been met by laying out two separate proofs. Fermat, however, considers that both concepts are implicit in his description. He also leaves out explicit instructions on how to construct the line NI_1.

Fermat's correspondents, who had learned their math catch-as-catch-can, and who usually had little experience using algebra, had never seen anything like his system before.

When Fermat began to show that one could describe more complicated curves with methods that represented such a radical departure from the historical norm, his readers understandingly found his explanations difficult.

Fermat's challenges were sometimes so difficult that his correspondents accused him of posing an unsolvable problem, and Fermat's habit of sending solutions afterward without sufficiently documenting his method must have been infuriating. Most of Fermat's correspondents who maintained a long-term relationship experienced the frustration of dealing with his idiosyncratic personality. Some eventually broke off communication with him altogether.

For the moment, though, the members of the Paris math community were patient, as they were hungry for his discoveries. And Fermat kept sending them.

Fermat's *Introduction to Plane and Solid Loci*

Fermat took his first steps toward creating an algebraic approach to geometry when he attempted to restore the *Plane Loci* of Apollonius. The *Plane Loci* was a collection of statements, ideas, and theorems about the geometry of straight lines and circles. But it lacked good demonstrations of how Apollonius had arrived at such knowledge. With algebra, Fermat tried to find a modern explanation for the truths Apollonius had found.

As Fermat worked his way through the book,

he gradually realized that all of the constructions Apollonius mentioned could be described with equations in two unknowns. (Their modern expressions in general form are the line $y = mx + b$ and the circle $x^2 + y^2 = r^2$. In each case the unknowns are the variables x, y; we call them indeterminate equations because there is an infinite set of values for x and y that make the equation true, rather than just a single answer.) Once Fermat realized this fact about lines and circles, it inspired him to examine the curves in another book by Apollonius called *Conics*. He discovered that these curves—parabolas, ellipses, and hyperbolas—could be described by the same family of indeterminate equations.

This realization was one of the moments separating ancient from modern mathematics. The Greeks thought of their curves physically, constructing lines and circles with compass and straightedge. When it came to the curves in the *Conics*, Apollonius had realized they could all be created by slicing a plane through a double cone at various angles—a famous insight that guaranteed his place in history, but a conception that still relied on a physical construction. To bring these curves together again in a completely different way—this time using the concept of an equation family—was no less an insight, and it gave Fermat a more powerful set of tools for exploring those curves than Apollonius had.

Fermat divided all first- and second-degree equations into seven groups, each of which had a

particular form and type of curve associated with
it. Here are Fermat's names and forms as we would
write them, with x, y representing variables and A,
B representing known values:

1) $Ax = By$ line
2) $xy = B$ hyperbola
3) $x^2 \mp xy = Ay^2$ line(s)
4) $x^2 = Ay$ parabola
5) $B^2 - x^2 = y^2$ circle
6) $B^2 - x^2 = Ay^2$ ellipse
7) $B^2 + x^2 = Ay^2$ hyperbola

Over the course of the *Introduction*, Fermat
used a graphic system to illustrate his curves, but it
differs substantially from the system we use today
in two respects. First, Fermat does not plot his
curves with a dual-axis coordinate system. Instead,
he employs a circular horizontal axis, which curves,
intersecting at some distance from the origin. The
other points on the curve—that is, those we would
say possess a non-zero y-coordinate—he imagines
resting above the axis at the end of a line segment
of varying length that extends from a point along
the axis at a certain angle. This angle is usually, but
not always, a right angle. In retrospect, it all seems
much more complicated than simply employing a
system of two axes, but in this regard Fermat was
proceeding from ideas of geometric construction
he had learned from Viète. As Fermat found the
system comprehensible, he did not see any need to
change it.

The second major difference between Fermat's system and ours is that he only considered positive values in his answers. His solution sets, therefore, only provide what we would consider to be the "top half" of curves, which lie within the first quadrant of the modern coordinate system. The addition of a y-axis and the possibility of negative values would be improvements others would make.

four

A Stream From the South

Though the Parisian circle included some of the age's finest mathematical minds, Fermat's problems frequently left them stumped. Their techniques represented the best that traditional math had to offer at the time, but Fermat had surpassed the tradition by this point, and they knew it. Within a few months, Roberval and Mersenne were begging Fermat for a general description of his methods and for permission to publish those methods.

Fermat gradually complied with the first part of their request. But while he proved willing to provide the Parisians with a reasonably good account of his achievements and methods, he had different feelings altogether when it came to publishing his works. Fermat clearly enjoyed the praise he was receiving for his mathematical efforts, but he was an amateur and wished to keep it that way.

Roberval and others offered to expand Fermat's notoriously terse explanations so that the path he followed from the start of a proof to its conclusion would not demand such numerous and long mental jumps on the part of his readers. Fermat refused. He was unwilling even to have his papers published anonymously.

Perhaps the Parisians should have expected Fermat's indifference to fame from the beginning. When he told them of his restoration of Apollonius's *Plane Loci*, he informed them that he would have to ask a friend to send it to them from Bordeaux. After writing it the decade before, he had left it there, neglecting to make any other copies—even for himself. Fermat did not see the need to keep his papers organized; he was a hobbyist and was not trying to write for fame or posterity. What he needed for himself he kept in his head.

Sometimes Fermat would forward an out-of-date copy of what he had kept of his methods, even though his thinking had gone far beyond what the version he sent indicated. Old, sketchy, and incomplete versions of his problem-solving techniques circulated through Paris. It was hard to get a clear perspective on everything Fermat had discovered and when.

Fermat's paper *Method of Finding Maxima and Minima and Tangents to Curved Lines* is a good example of how difficult it could be to follow his work. The paper contained some of Fermat's most important contributions to the foundations of later branches of mathematics, though the theoretical justifications behind the work were not as clear as had been the case with his *Introduction*. The justifications did exist; he had possessed them since doing his initial work on the paper in Bordeaux around 1629. But as is so common with Fermat, possessing and sharing were different matters. He sent the initial paper to Paris in 1636. Over the next few years, largely in response to criticisms of its lack of theoretical underpinnings, Fermat was eventually drawn to flesh it out. With the benefit of hindsight, his discoveries are more easily appreciated.

In the *Introduction*, Fermat had demonstrated that an equation could contain all the information necessary for a curve's construction. The method of tangents from *Maxima and Minima* showed that an equation could reveal still more: where a curve's highest and lowest points were; the nature of the line that grazed its edge. But there was yet another piece of knowledge that equations might reveal, with the right approach. Though it took Fermat most of his life to explore fully, he was already on the trail of this piece of knowledge by the time he left Bordeaux.

The piece of knowledge in question was how to find the area of the space between the curve and the axis. Finding the area of a regular shape like a rectangle had long been a simple matter of multiplying its length by its width. An irregular polygon with odd-shaped sides might take a bit longer, but with a bit of imagination the polygon could be broken up into squares and right triangles, whose total area could be computed without much difficulty.

Finding the area of a shape with a curved side was not so easy. The entire concept of area, in Fermat's time as in the present day, is thought of in terms of squares—square inches, square feet, square meters. When confronted by a shape with a curved side, then, mathematicians from antiquity until Fermat's day had done their best to break up these unusual shapes into tinier and tinier pieces, something like what could be done with the aforementioned irregular polygon.

With a parabola curving up and to the right of the axis, for example, a geometer might try to fill up the space beneath it with rectangles, each one a single unit in width but with different heights, each stretching up from the axis to a different point on the curve. By finding the areas of each of these rectangles and adding them all up, the geometer could find an approximate value for the total area between the axis and the parabola. It was an extremely laborious process. Yet it could never yield an exact answer, because regardless of how narrow the geometer made the rectangles, they could touch the curve at only one point. This meant that there was a space above each rectangle and below the curve whose area was not included in the total area.

Fermat eventually found a way to imagine the size of the rectangles becoming infinitely small, as well as formulae that could tell him what these rectangles' total area approached. This allowed the area under the curve to be calculated with great accuracy.

All this was news to the Parisians, though it is familiar to many students of math today. There is a good reason for this. The discoveries Fermat was making in the early years of his career about four centuries ago form the basis for the tools used by algebra and calculus students today to describe curves and lines with variables and exponents.

During his time, however, Fermat's unwillingness to present ideas clearly and publicly drew criticism. But not all Fermat's difficulties with his peers stemmed from his desire for relative anonymity or from the intense demands he placed on his readers. A more important factor may have been his naïveté. He apparently did not recognize that, out of jealousy or impatience with his indirect style of communication, others might be spurred to attack him personally. Over the years, Fermat received his share of angry and insulting letters from other mathematicians and scientists. His responses, though, were unfailingly polite, dignified, and restrained. He was unwilling to stoop to personal insult.

Some of Fermat's work seemed confusing or even irrelevant to his contemporaries. But there was a period in his life when he was at the center of one of the most heated mathematical debates in Europe, and with one of the most influential thinkers in history. Shortly after Fermat's introduction to the Paris scientific circle, he would have his great trial by fire—an experience that would illuminate his already brilliant reputation but also leave it scorched.

Maxima and Minima

Fermat begins *Maxima and Minima* with an example of how to maximize the size of a particular rectangle. Specifically, given a line segment of length B, at what point along B's length should the segment be bent into a 90-degree angle so that the resulting two smaller segments contain a rectangle of maximum possible size?

Fermat calls one part of the line segment A, the other $B - A$. The area of the rectangle, length multiplied by width, is thus $A*(B - A)$ or $AB - A^2$.

He then imagines another rectangle that has the value of A replaced by $A + E$, for which E is some unknown interval. The second rectangle's area is then $(A + E)(B - A - E)$, or $AB + EB - A^2 - 2AE - E^2$.

If the two rectangles are both of maximum size, Fermat reasons, then their areas should be equivalent:

$$AB - A^2 = AB + EB - A^2 - 2AE - E^2$$

By subtracting $(AB - A^2)$ from both sides and dividing through by E, he finds that $B = 2A + E$, but if E is reduced to zero, $B = 2A$, and the rectangle is therefore a square.

The findings from this example are trivial in and of themselves; it is the method that is significant. Fermat says:

> Let A be the unknown of the problem. . . . We shall now replace the original unknown A by A + E and we shall express thus the maximum or minimum quantity in terms of A and E . . . , equate the two expressions of the maximum or minimum quantity and take out the common terms . . . [and] divide all terms by E. . . . The solution of this last equation will yield the value of A, which will lead to the maximum or minimum, by using again the original expression.

In this instance, Fermat was employing a concept called *adequality*, which roughly means to consider two unequal values as being approximately equal (*ad-equal*). The Greeks considered this idea useful for visualizing problems, but not rigorous enough for proofs; Fermat showed it could be both. The same method, he found, can be used to find extreme points on curves.

In *Maxima and Minima*, Fermat also put forth a method for finding tangents. For the Greeks, a tangent was a straight line that touched a curve at a single point (without intersecting it). This purely geometric definition had grown, by Fermat's time, into an important dynamic concept. Galileo had recognized that an object thrown, launched, or fired from a gun would always follow a path described by a parabola. Moreover, at any particular moment, the object's direction would be described by a tangent to that curve. But seventeenth-century mathematicians were still hard put to answer, in useful ways, questions such as what the object's velocity would be at a particular moment.

Since his days in Bordeaux, Fermat had been considering the math that would lead to some of these answers. As usual, however, he was not concerned with how a scientist might use his discoveries to explain the physical world. Rather, he was captivated by the abstract beauty of tangents and hoped simply to solve mathematical problems for their own sake. Many scientists of the day spent considerable time finding tangents to curves, but it is Fermat's approach—which he discovered in Bordeaux in 1629 and included in *Maxima and Minima*—that most closely resembles the one we use today.

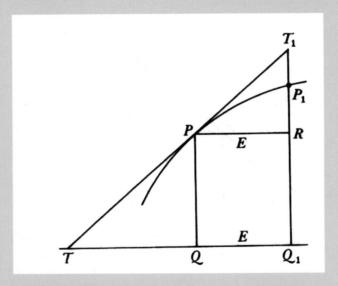

Fermat inscribes a curve above his axis and draws a tangent from T on the line to P on the curve. PQ is a line segment from the axis to the tangent; the unknown length TQ along the axis, called the subtangent, completes the triangle TPQ. Once Fermat shows us how to find the length of TQ, we can use our knowledge of triangles to describe exactly the line of which TP is a part.

By extending TQ along the axis to point Q_1 by a very small interval, called E, and also extending the tangent up to point T_1, we now have two triangles, TPQ and PRT_1, that have the same proportions (mathematicians call such triangles *similar*). From this proportionality, we can say

$$TQ{:}PQ = E{:}T_1R$$

But because of the very small intervals we are dealing with here,

$$TQ{:}PQ = E{:}\ (P_1Q_1 - QP)$$

In modern terms, we can call PQ a function $f(x)$, and rewrite the above as

$$TQ{:}f(x) = E{:}\ [f(x + E) - f(x)]\quad or\quad TQ = E * f(x)\ /$$
$$[f(x + E) - f(x)]$$

As Fermat treated $f(x)$, he could then remove the E term from both numerator and denominator and obtain TQ.

five
A Comedy of Errors

By the fall of 1637, Fermat's life appeared to be going quite well. He had a solid, lucrative legal career and enjoyed a stable marriage. In addition, he had been discovered by a group of serious mathematicians in Paris, including one—Mersenne—who could serve as his bridge to the rest of Europe. If Fermat remained isolated geographically from mainstream mathematicians, he was no longer cut off from them intellectually.

But if some of the mainstream had begun to discover him, Fermat himself had yet to discover a true mathematical peer. His correspondents included several who—when Fermat cared to be explicit about his new methods—could follow and appreciate his advances, but none who were his equal in mathematical creativity and inventiveness. When he did finally come across someone who could engage him on his own level, however, the exchange was far from pleasant.

That someone was René Descartes (day-*cart*), an ambitious philosopher whose writings on deductive reasoning would profoundly influence the subsequent development of

René Descartes

European thought. Descartes was one of the main architects of rationalism, a movement that sought solutions to human problems chiefly through the application of reason and that demanded demonstrable proofs to support claims, rejecting appeals to religious authority.

As mathematics was an inherently rational activity, Descartes considered it to be a discipline that could handily demonstrate the advantages of his new system of thought, which he was outlining in his *Discourse on Method.* Unlike Fermat's *Introduction to Plane and Solid Loci,* the *Discourse* was not a book primarily for mathematicians. Descartes intended it for anyone who wanted to use reason as a problem-solving tool.

Descartes envisioned his advancements in math as a vital aspect of his claim to fame as a thinker. As an example of mathematics' utility, he dedicated a large section of the *Discourse* to his newly devised system of analytic geometry. Unbeknownst to Descartes, Fermat had created his own very similar system of analytic geometry years before, during his time in Bordeaux. And at the same time that Descartes was circulating the manuscript of the *Discourse* among his friends in Paris, Mersenne received a copy of Fermat's *Introduction,* a work that clearly showed methods of plotting geometric figures on an axis and analyzing them with algebra.

The simultaneous discovery of a new scientific truth or technique has occurred often in history, and it does not always result in the eruption of a heated debate over who deserves credit for the advance. Sometimes the two parties share the credit, or the scientific community judges one person's efforts clearly superior. No such simple ending came of the dispute that arose between Fermat, Descartes, and their respective

DISCOURS
DE LA METHODE

Pour bien conduire fa raifon, & chercher
la verité dans les fciences.

P L U S

LA DIOPTRIQVE.

LES METEORES.

E T

LA GEOMETRIE.

Qui font des effais de cete METHODE.

A LEYDE
De l'Imprimerie de IAN MAIRE.
cIↃIↃc XXXVII.
Auec Priuilege.

The frontispiece of the original version of Descartes's *Discourse on
the Method for Conducting One's Reason Well and for Seeking Truth in the
Sciences (Courtesy of Lebrecht Music and Arts Photo Library/Alamy)*

groups of supporters over the creation of analytic geometry. But by the time it subsided, the controversy would carry Fermat's name across Europe.

About a year before Pierre de Carcavi went to Paris and introduced Fermat's work to the Parisian intellectuals, Fermat's friend and fellow Viète disciple Jean Beaugrand left the provinces for a job as royal secretary in the capital. Although the two evidently never saw each other after 1635, Fermat and Beaugrand maintained cordial relations until the latter's death in 1640.

It is unlikely that Fermat would have continued the friendship had he seen how Beaugrand was behaving in Paris. In letters to other scientists, Beaugrand had passed off ideas Fermat had shown him as his own.

In addition to being unscrupulous, Beaugrand was evidently rude. He managed to offend nearly all of his scientific acquaintances in Paris. When, in early 1636, he published a paper claiming that an object's weight varies as its distance from the center of the earth changes, few in Paris were in the mood to support him. (Scientists of today agree that this is true.) Of his many critics, one of the most vocal was the well-respected Descartes, who was also a correspondent of Mersenne's.

The scientific community mocked Beaugrand, and the insults stung so deeply that he began attacking Descartes in anonymous pamphlets (which at the time was a common way to spread rumors). Bad blood seethed between the two men.

A few of Beaugrand's friends in Italy supported his views, and Mersenne needed someone to help resolve the situation. It was at this time that Carcavi had arrived in Paris and had begun to tell Mersenne of his talented friend in Toulouse. Far from the emotionally charged atmosphere in Paris, Fermat had no idea of the trouble he was about to face when he received his first letter from Mersenne, asking his opinion on Beaugrand's views.

With regard to the quality of Beaugrand's science, Fermat came down on the side of his old friend. But as he had also enclosed much of his own work in the same letter as his answer, the majority of the Parisian circle quickly became more interested in discussing Fermat's accomplishments in math, something Fermat was only too happy to oblige. In any case, Beaugrand's subject—called geostatics—was quickly being replaced with more advanced conceptions of gravity (which didn't contradict Beaugrand's ideas, but expanded on the same theories more fully and eloquently). Scientists like Galileo and Descartes were shifting their focus to these other ideas, and discussion of Beaugrand's paper had ended completely by December of 1636.

Fermat had avoided trouble, but only for the moment. Beaugrand was still itching for revenge, and he wanted to mobilize his friends against Descartes. The opportunity came just a few months later.

In early 1637, Descartes was preparing a paper of his own, the *Dioptrics*, for publication. It dealt with the reflection and refraction of light. Mersenne, concerned as always with ensuring that good science would become available as soon as possible, did what he could to ensure that Descartes' work would be published: he spoke to the official whose job it was to obtain licenses for printing new books. That man was the royal secretary, Jean Beaugrand.

Beaugrand had Descartes right where he wanted him. Abusing his official position, he obtained a manuscript copy of the *Dioptrics* and sent it to Fermat for comment.

Fermat still had no awareness of the feud that was roiling in Paris, but he was nearly saved again from trouble because he was so busy communicating with other mathematicians. His analytic geometry was making an impression, and he was spending what spare time he had answering questions about its various capabilities. These questions were numerous, largely because Fermat—as was becoming his habit—did not provide any correspondent with a full explanation of his discoveries but rather spread key elements of his ideas among several letters. Occupied with showing Paris his discoveries in this piecemeal fashion, Fermat put Descartes' paper in the stack on his desk, where it remained for a couple of weeks.

In the meantime, Mersenne had discovered Beaugrand's dubious behavior. Mersenne knew that Descartes, who was capable of blistering anger when provoked, would not be happy if he learned that Mersenne had played any part in allowing his enemies to see the *Dioptrics* before publication. Mersenne quickly dispatched a letter requesting that Fermat send any comments about Descartes' manuscript directly to him, rather than to Beaugrand.

Always happy to oblige Mersenne, Fermat immediately replied with a response for Descartes. He said that he had not had time to read the paper closely enough to address each point individually, especially as the subject matter was "thorny." But, Fermat said, "when I consider that the search for the truth is always praiseworthy and that we often find what we seek by groping about in the shadows, I felt you would not think ill of me if on this subject I tried to outline my ideas for you, which, since they are obscure and halting, I will perhaps clarify some other time."

Fermat had not meant to be insulting, but when he implied that the author of the *Dioptrics* had come upon his discovery by "groping about in the shadows" rather than by intelligence and observation, Mersenne knew Descartes would be annoyed. He decided to hold on to the letter rather than forward it immediately.

Meanwhile, Descartes was preparing an even more ambitious work, the *Discourse on Method*, which was intended to make his name as one of the great thinkers of all time. A fundamental part of the *Discourse* was his method of analyzing geometric figures with algebra.

More anonymous pamphlets appeared before the publication of the *Discourse*, criticizing its author for his system of thought. Descartes realized that Mersenne had allowed the wrong people to see his paper. He began to suspect that a conspiracy had formed to destroy his reputation.

When Descartes asked Mersenne to forward any criticisms he had collected of the *Dioptrics*, Mersenne finally sent along Fermat's letter. Descartes recognized the name from one of Fermat's less sophisticated mathematical letters, which had been circulating among members of Mersenne's circle. That letter, which concerned Fermat's solution to a single brief problem from Apollonius, gave Descartes no reason to believe this mere magistrate from the provinces had an intellect to be reckoned with. Through Mersenne, Descartes sent a rather condescending letter to Fermat in October, requesting that Fermat reread the *Dioptrics* to be sure he had understood it.

Within a few weeks Fermat had responded. He sent Mersenne—to be forwarded to Descartes—his *Introduction to Plane and Solid Loci*, along with his related work on maxima and minima. Fermat also hinted that he had been surprised the *Discourse* contained nothing about the determination of extreme points (maxima and minima).

Upon seeing these far more sophisticated works in December, Descartes realized that he was dealing with a mathematical talent of the first rank—or Fermat had obtained an advance copy of the *Discourse* and plagiarized it. In either case, Descartes concluded that Fermat must be part of the conspiracy that was seeking to destroy his reputation. He lashed out at Fermat.

Descartes' maintained a certain condescension toward Fermat, but his arguments grew more pointed as his defense mounted. The *Discourse*, he insisted, did contain a method of determining maxima and minima; it was implicit in his method of determining tangents to a curve. Moreover, Fermat's methods were unacceptable because they relied on making a counterfactual assumption at a certain point. This was Fermat's use of adequality, which, though it gave the correct answer in the example Fermat had provided, did not extend to a general class of problems. An adequality, or "approximate equality," is an equation obtained by making a deliberately false assumption about a system, manipulating it algebraically, and then removing the false assumption and extending the conclusion to the general case.

As a method, Descartes said, it was insufficient. Fermat had made the same sort of criticism of the *Dioptrics*, and Descartes intentionally threw it back in his face. "His rule (i.e. the one he takes pride in having found) is such that with no work and by accident one can easily fall into the path that one must take to find it," Descartes wrote of

Fermat's method. No work and by accident—this was as bad as saying that Fermat had been groping about in the shadows himself.

When Descartes dispatched this vitriolic letter to Mersenne on January 18, 1638, the friar did not send it along to Fermat immediately. No evidence remains as to why. Perhaps, in the face of such withering criticism from a man as eminent as Descartes, Mersenne may have questioned his initially favorable impression of Fermat's abilities. In any case, his next action only exacerbated the growing conflict. He sent the reply to Gilles Roberval and Etienne Pascal, who were also enemies of Descartes. They began to attack Descartes' methods themselves, exchanging several letters with him via Mersenne through the spring of 1638. Descartes finally asked Mersenne to bring in a third party, Girard Desargues, to determine whether Fermat was indeed proceeding from a firm mathematical basis.

It is important to recall that over the course of this entire exchange, Fermat and Descartes had never once corresponded directly. Rather, they had sent letters through others. The two greatest mathematical minds of the age had been drawn into a bitter exchange by intermediaries with their own agendas.

Though it became clear that Descartes was attacking his mathematical abilities, Fermat never said anything even mildly insulting to Descartes. He calmly explained his concepts more clearly than he had in the letters that had been circulating through Paris. Desargues, meanwhile, came to the conclusion that Fermat had merely failed to word his method clearly enough in the letter Descartes had seen, making it seem less than general. But in fact the method was general.

"M. des Cartes is right," Desargues said, "and M. Fermat is not wrong."

By July the dispute had run its course, and Descartes conceded in a private letter that Fermat had indeed found a workable method for finding the maxima and minima of curves. Yet Descartes continued to disparage Fermat in letters and conversations with other scientists. As he was a highly respected scientist, Descartes' words carried considerable weight. Fermat's name was tarnished.

Finding maximum and minimum points of curves was obviously not the only thing Fermat's analytic geometry could accomplish, and the underlying issue for Descartes was far greater than a single mathematical technique. Ultimately history would give Descartes most of the credit for the creation of analytic geometry, even though Fermat's analytic geometry was also effective. Fermat's system had some advantages, but it was not laid out as clearly as Descartes'—and as Descartes published his results in the context of a larger philosophical system, his achievement easily won him the fame he had desired. The two-axial system used today for plotting points in algebraic geometry is still called the Cartesian coordinate system in his memory.

For his part, Fermat had never seemed interested in winning fame. Yet the scientific community had come to know him as a mathematician who could spar with Descartes on equal terms, even if such conflict was contrary to his nature. Distressed by the shrill tone of the argument over analytic geometry, Fermat turned back to his own investigations. He even picked up the *Discourse on Method* and began to improve his own system based on the advantages Descartes' system could offer.

Within two years, Descartes had lost interest in mathematics altogether, concentrating his efforts on larger issues in philosophy. Fermat, however, had only improved his mathematical problem-solving skills as a result of the skirmish, and he was ready to do more.

Adequality had been just one of his techniques for solving problems, but the need to justify himself to Descartes and the world repeatedly focused his mind upon it. The concept had led him to a way to find maxima and minima. Where else could it take him? He let the question sit in his mind for a while.

six

The Lure of Pure Numbers

The dispute over analytic geometry was a storm that had swirled around Fermat and engulfed some of his finest achievements. But in the storm's eye, Fermat calmly remained focused on his first love: pure numbers and what they could show him. His mind drifted back to the reasons he had found the world of mathematics so attractive to begin with.

Geometry's orderly shapes and smooth curves had been a great pleasure, and he had played an important part in restoring the understanding of ancient mathematicians. Algebra had illuminated the landscape more brightly than the Greeks' physical constructions and skill with proportions ever had. But what were those geometric figures really, and what were those equations that described them? They were signposts to the patterns the Pythagoreans had imagined, but they were not the patterns themselves. They had provided the pathway to understanding numbers, but they were not the pathway's end. It was time to travel further, back to the source.

The source for Fermat is today called number theory—the study of the properties of numbers themselves, independent of what those numbers might represent, such as the length of a circle's radius or the area of a polygon. Though number theory was Fermat's great passion, the mathematical world would not learn to appreciate it until more than a century after his death. Number theory today is nearly as odd and apparently irrelevant a subject as it was in Fermat's day, and in the time of Pythagoras. It is a subject largely unconcerned with its own practical application, though in recent decades it has found some important uses in data encryption. It is a subject that has only limited utility even for elucidating other areas of mathematics. Nonetheless, it is a discipline with sufficient abstract beauty to attract the attention of many of history's best conceptual thinkers, including Carl Friedrich Gauss, considered by some to be the greatest mathematician who ever lived.

Number theorists look for patterns that exist within the integers (. . . -2, -1, 0, 1, 2 . . .). The most famous kinds of patterns they seek to understand are those relating to prime numbers—those numbers that are greater than 1 and are divisible only by 1 and themselves, (such as 2, 3, 5, 7, 11, 13, and 17). Prime numbers have fascinated mathematicians since Pythagoras, and mysteries about them still abound. One mystery regards how often primes crop up in the sea of composite numbers, which have other factors besides 1 and themselves; 2,500 years after the Pythagoreans started asking such questions, no one knows the precise answer.

In some ways, the questions number theorists ask, and

The abstract beauty of number theory was attractive to many famous mathematicians, including Carl Friedrich Gauss.

the patterns they find, seem like an elaborate game played for its own sake. For example, number theorists take great joy in finding what they term perfect numbers. Numbers are said to be perfect if, when all the positive whole numbers

Pythagoras (center, writing)

that can be divided evenly into them are added up, the sum is the number itself. The number 28 is a perfect number:

$$1 + 2 + 4 + 7 + 14 = 28$$

Early on, Mersenne had heard of Fermat's talent with numbers themselves, when Carcavi had mentioned this talent in regard to issues in music theory. Fermat had tried, without success, to interest the Paris group in his work on number theory; the group instead focused on his work with analytic geometry. But Fermat kept trying, and eventually Mersenne passed along his work to Bernard Frenicle de Bessy, a young Parisian who was quite talented at manipulating large numbers. Although Frenicle was a few years younger than Fermat and not nearly as gifted in math, he was probably the closest thing to a kindred spirit Fermat would ever encounter.

In March 1640, Frenicle wrote to Fermat through Mersenne. He challenged the magistrate from Toulouse to find a perfect number of twenty or more digits. Fermat, who was by this time notorious for simply posing problems to his correspondents, must have been pleased to receive Frenicle's letter. Fermat slightly misread the letter's wording, however, and his response reveals something of the way he had been thinking about number patterns. Specifically, Frenicle had asked Fermat to find a perfect number of twenty digits "or the next one following," which Fermat took to mean a number of exactly twenty or twenty-one digits. Fermat replied: "I have several shortcuts for finding perfect numbers, and I can say in advance that there is none of twenty digits, nor any of twenty-one digits."

Frenicle already knew this, and he understood the source of confusion. But Fermat's mention of "shortcuts" piqued the younger man's curiosity. Through Mersenne, the two began a correspondence that revealed how much Fermat had been thinking about prime numbers, for primes were the key to the shortcuts of which he spoke.

Around June, Fermat wrote to Frenicle that he had three propositions regarding patterns in perfect numbers (which Euclid had proved all have the form $2^{2n+1} - 2^n$). Fermat said he hoped "to raise a great structure" on these propositions, all of which concern numbers of the form $2^n - 1$. All prime numbers of this form, Fermat said, generated perfect numbers—if one knew how they operated. The key, Fermat said, was looking at the primality of the exponent n in the expression $2^n - 1$. Fermat's three propositions were as follows:

1. If n is composite (that is, not prime), then $2^n - 1$ is composite as well.
2. If n is prime, then n evenly divides the number $2^{n-1} - 1$.
3. If n is prime, then the only possible divisors of $2^n - 1$ are of the form $k(2n) + 1$.

This third proposition was what had allowed Fermat to answer Frenicle's question so easily. For Fermat knew that the only possible value of n in $2^{2n+1} - 2^n$ that yielded a perfect number of either twenty or twenty-one digits was 37, or $2^{37} - 1$. (The numbers 32 to 36 are not prime, nor are 38 to 40; even without working out the precise value of $2^{2n+1} - 2^n$ for $n = 31$ or 41, the two closest primes, Fermat would have realized quickly they did not possess the correct number of digits.) The only remaining question was, does $2^{37} - 1$ yield a prime number?

He knew $n = 37$; all he had to do was try different values for k, which did not take him long. Fermat only had to go as far as $k = 3$ to determine that $3(2 \times 37) - 1 = 223$, which divides $2^{37} - 1$ evenly. Therefore, $2^{37} - 1$ was not prime, and $2^{2n+1} - 2^n$ for $n = 37$ was not perfect.

What he did not reveal to Frenicle was how he had determined these rules. But he went on to say, "From these shortcuts, I already see a great many others emerging, and for me it is like seeing a great light." And the following October, he revealed one of these other propositions, to which the first two propositions he mentioned in June were merely corollaries. This observation was what has come to be known as Fermat's Theorem. In modern language it would be expressed as follows:

> If b is any natural number and p is a prime number that does not divide evenly into b, then p has to divide evenly into $b^p - b$.

Though this statement also is sometimes referred to as Fermat's Little Theorem, that name is used merely to distinguish it from his legendary "Last Theorem," another of Fermat's famous contributions to the world of mathematics. But his Little Theorem is anything but little with regard to its utility, for the discovery has proved key to the secure encryption of data on the Internet. Fermat, of course, lived centuries ahead of the digital age, and his interest in statements like the Little Theorem concerned their ability to reveal patterns among the prime numbers.

An infinite number of primes exist. Because they crop up

unpredictably on the number line, however, distinguishing them from the composites has bedeviled mathematicians throughout history. One method is the famous "Sieve of Eratosthenes," a process of elimination named for the third-century BCE. mathematician who discovered it. To find all prime numbers up to 100, for example, the sieve would work as follows: Write out all the numbers from 1 to 100. Circle the number 2, then cross out all the numbers divisible by 2. Circle the next lowest unmarked number (3), and cross out all the numbers divisible by it. Repeat the process for all the remaining numbers. The circled numbers will all be primes.

The sieve may be simple and effective, but for very large numbers it works too slowly to be of use. Even sophisticated modern computers have trouble running the sieve with numbers of many thousands of digits. Fermat did not

have access to computer technology, but the problem for him was the same one that exists today: how can patterns be found in a sequence of numbers that seems to be patternless? Answering such questions requires keen mathematical intuition, but Fermat had formidable intuitive powers.

Eratosthenes, the third-century BCE mathematician who discovered a method to determine prime numbers.

His intuition led him to seek a way to see straight through to a number's primality without bothering to find its factors in the first place. The Sieve of Eratosthenes, for all its usefulness as a filter, also provides the user with information about the factors of every composite number in the list— needless extra detail. But Fermat's Little Theorem still comes in handy nearly four centuries after he conceived it. In fact, the best computer algorithms used to find primes today are all descended from it.

If, for example, $b = 2$ and $p = 5$, the theorem predicts correctly that 5 is a factor of $2^5 - 2$, or 30. If we wish to find a large prime number, Fermat's Little Theorem opens a door to a method.

Because the sieve eventually becomes too unwieldy and time-consuming for finding large primes, one approach is to choose a number p thought to be prime, work out the value of $b^p - b$, and divide the result by p. If the solution leaves a remainder of zero, then p is likely prime.

One problem exists, however. Fermat's Little Theorem also holds true for some composite (that is, non-prime) values of p. There are not many of these composites to worry about; the smallest is the number 341, which is the product of 11 and 31, and there are only 244 other such composites between zero and a million. But while they are rarely occurring, there are nonetheless an infinite number of them, and so a bit more work is necessary to turn Fermat's Little Theorem into a reliable source of large primes.

In 1986, additional tests were developed that, combined with Fermat's Little Theorem, allow a supercomputer to test even a fifty-digit number for primality in a few seconds.

Many wonder where Fermat came upon his ideas about

the properties of prime numbers. Fermat's thinking about analytic geometry is difficult to trace, but he had many correspondents who asked him repeatedly about his work, and he also became embroiled in a controversy that eventually forced him to reveal his methods. Since far fewer were interested in his work in number theory, he left even fewer clues behind. Number theory has been called a game of inspiration, and the sources of Fermat's can only be outlined tentatively.

Prime Numbers and Fermat's Little Theorem Today

Number theory is concerned with making discoveries for their own sake, with little or no thought for how these discoveries may be applied. But while tools must sometimes wait a long time for their tasks, utility does have a way of creeping in eventually.

One such application stems from the fact that a functioning supercomputer can quickly test large numbers for primality. Thus, an excellent way to ensure that a new supercomputer is working properly is to give it a really large number already known to be prime, and ask the machine to determine for itself whether or not the number is prime. Manufacturers frequently run this diagnostic test on their latest creations.

A more important use, though, is for the encryption of data, particularly on the Internet. People link their computers using wires and transmissions that are vulnerable to monitoring by others—not a pleasant scenario if the data being sent contains private information such as bank account numbers and the secret codes necessary to access them. Keeping this information secure is a matter of critical importance, especially today when many people make large financial transactions via the Internet. Fortunately, a high level of security is possible because of some key properties of prime numbers, combined with Fermat's Little Theorem.

A supercomputer can test two different fifty-digit numbers for primality in less time than it would take an average person to read this page. But if those two numbers are multiplied to obtain a third number of about one hundred digits, and the supercomputer is asked to find the two original numbers, the task would require many years.

The reason for this disparity is that, while mathematicians like Fermat have found shortcuts for determining primality, no such shortcuts have been discovered for finding factors. This has allowed programmers to create software with the aid of which a coded message can be sent over insecure lines without concern for who else might see it.

This software provides users with two prime numbers of, say, seventy-five digits; multiplying these two numbers together provides a number of approximately 150 digits that is a vital part of the encryption key. Decoding the message, however, requires the recipient to factor the 150-digit number. Fermat's Little Theorem is involved in both sides of the process. Because it is so difficult to factor such a large number, even the best computers are helpless to decode messages sent with such encryption methods. In theory, it is possible to do so. But for practical purposes, no one can.

seven

The Shape
of Numbers

"I am led to believe that you have recently devised for yourself some sort of special analysis for prob- ing the most hidden secrets of numbers."

With these words in late 1641, Bernard Frenicle expressed his astonishment at Fermat's recent letters to him, letters that contained problems even more challenging than usual. Fermat was rapidly moving past the subject of prime num- bers and was now considering another aspect of numbers that, at first glance, might appear to represent a step back to his days of algebra and geometry. His letters were now full of observations about triangles and squares, and some- times other shapes. But this time, Fermat was not talking about constructions made of lines and points, at least not directly. This time, he was building numbers themselves into shapes.

Though a number has always meant a group of units, Fermat began to understand these units as movable objects he could rearrange into patterns in his mind, as though each were a toy block he could stack and manipulate. Starting from a few familiar objects, he began finding ways to break these shapes apart and put them together into others—triangles into squares, squares into triangles, larger squares into multiple smaller ones. Before long he was lining them up, breaking them down, and transforming these shapes in ways even Diophantus had not imagined, though it was Diophantus that got him started.

The *Arithmetica* was full of problems that concerned numbers imagined as polygonal shapes, in which the arrangement of the units was paramount but the distinction between units of length and area blurred. Diophantus had, for example, challenged his readers to find four numbers such that the product of any two of them, plus one unit, is a square. He also challenged them to find a right triangle such that the number of units in its area, when added to the number of units in the length of one smaller side, equals a given number.

After nearly two decades with the *Arithmetica*, Fermat was an old hand at problems like these. He knew every arithmetic trick Diophantus had used. Diophantus had not provided instructions for when to use a particular technique, but Fermat intuitively grasped which tools might be effective in solving specific problems. Still, he recognized that being able to puzzle through problems on a case-by-case basis is different from having a universal method for solving them.

Fermat saw Diophantus searching for right triangles one at a time, as though exploring a dark forest with a single candle rather than a bright lantern that could illuminate the entire landscape. He had been good at breaking up certain problems into two equations, which he could then solve—sometimes. Other times, using Diophantus's techniques yielded a negative answer, which as a length or area in the physical world was meaningless. Yet again, Fermat wondered about how to encapsulate a collection of tricks as a straightforward, universal method.

All the while, he was busy with other issues as well. The dispute with Descartes had taken its toll on him, and he kept returning to some of the ideas that he had been forced to explain in order to prove his ability to the doubters in Paris. Adequality, Fermat knew, was an idea whose roots extended far back into history. He had found it by immersing himself in Greek achievement, not by groping around blindly. Why should he have to justify himself if others did not look as closely as he had at antiquity?

But justifying himself is exactly what was demanded. Over the previous few years, he had done something he never cared to do: he sent letters outlining parts of the process by which he developed his methods of maxima and minima. He also outlined the idea of adequality, which made the process work: for a clearer perspective, add a bit of new information. Then work through the problem a second time and see what it reveals.

It might have been this focus that brought him to one of the next great mental leaps of his career, one that led him closer to the understanding of pure numbers he had pursued from the beginning. He came back to Diophantus and saw a way to add a bit of new information to the sets of equations the ancient Greek master had used to find right triangles. In so doing Fermat was able to turn Diophantus's candle into a lantern, illuminating a whole forest of right triangles.

Frenicle posed a problem to Fermat in 1641 that proved difficult for both men, but within two years Fermat had the answer. To find it, he had needed to forge a complete method from the tools he inherited from Diophantus, and this process marks Fermat's jump from ancient number theory to the modern form he created.

The problem was this: find a triangle in which the square of the difference of the two smaller sides exceeds twice the square of the smaller side by a square number. A square number is one that has an integer as its square root, for example 9 or 16. Mathematically represented, $(B-A)^2 - 2A^2 = y^2$ where the two shorter sides of the triangle from shortest to longest are A, B, and the squared number is y. From their experience with Diophantus, both Frenicle and Fermat had learned some tricks that worked in solving many problems of this type. One helpful trick was to set the triangle's two smaller sides as $B = x^2 + 2x$ and $A = 2x + 2$. They could then set up the problem as

$$(x^2 - 2x)^2 - 2(2x + 2)^2 = y^2 \quad \text{or} \quad x^4 - 12x^2 - 16x - 4 = y^2$$

This was a tough equation to solve, and they knew that getting rid of the higher-power variables would simplify it immensely. One way they knew of to do this would be to set y equal to a quadratic expression in x, or $y = Rx^2 + Sx + T$, with values of R, S, and T that will make the unwanted terms on both sides of the equation equal, allowing them to be canceled out.

In problems similar to the one above, this would be a

workable technique. In this particular case, though, the only appropriate values of R, S, and T lead to an answer of $x = -5/2$, meaning one side of the triangle would have a negative length—a geometric impossibility. Fermat's insight—which he probably had sometime in 1642—was to realize that, after obtaining the answer $x = -5/2$, one could substitute $x = (u - 5/2)$ into the original equation, and a new equation emerged:

$$u^4 - 10u^3 + (51/2)\, u^2 - (37/2)\, u + 1/16 = y^2.$$

From here, one could set $y = u^2 - 5u - \frac{1}{4}$ (again, so as to cancel the higher-degree terms on both sides) and obtain the value $u = 21$. So $x = (21 - 5/2) = 37/2$, a positive value that gave a triangle with sides of meaningful length. And if one wanted another triangle that fulfilled the conditions of the problem, all that was required was to use $(u + 37/2)$ instead of $(u - 5/2)$ and go through the process again, as often as necessary.

Clear parallels exist between Fermat's approach to his 1642 work in number theory and his approach to previous work with maxima and minima (which, in 1642, he was in the process of justifying to his critics in Paris). In both cases Fermat's method essentially relied on the idea of substituting $(x + m)$ for x into an unsolvable equation, working through it, and then simplifying it to remove the additional variable.

Fermat's method did not become clear until the late 1650s, when he described it in a letter to Father Jacques de Billy, a Jesuit math teacher. Billy, who eventually quoted Fermat at some length on the technique in one of his own papers, attested to its novelty. It is one of the few insights into Fermat's pathway into number theory that remains.

It was this method that in 1643 allowed Fermat to pose some apparently unsolvable problems to Frenicle and to another mathematician in the Paris circle, Pierre Brûlart. Both men were annoyed to distraction. Fermat eventually revealed one of the answers, though he kept the method to himself, as he always did. Despite the irritation of his correspondents, Fermat maintained his usual composure when he wrote Mersenne that August:

> One should propose the other problems for solution to those who say (as M. de Carcavi has written me) that I found my method of maxima and minima by accident. For perhaps they will not think that I found these problems [gropingly and by chance]. . . . I'd be obliged if you would soothe [Brûlart] on my behalf. Perhaps for his sake I will set down in writing my discoveries concerning Diophantus, where I have uncovered more than I could ever have promised myself.

Indeed he had, for by this discovery, Fermat put an end to the Diophantine tradition in finding special triangles one at a time. He had taken all of Diophantus's problems on the subject and found they could be solved with a single straightforward procedure.

Fermat frequently sent his correspondents what might be termed "challenge problems." While he posed a specific question and asked his correspondents to try their hand at solving it, in many cases he had deeper issues on his mind than the solution to a single problem. He did not spell those issues out, however. Rather, he was signaling that he had discovered something important, which he believed his correspondents would find for themselves if they looked at the problem closely enough.

Pierre de Fermat *(Courtesy of Mary Evans Picture Library/Alamy)*

Unfortunately, this purpose was invariably lost on his correspondents, even Frenicle. Certainly Frenicle had grown accustomed to the difficulty of Fermat's problems. But for all his talent, he was not as perceptive as Fermat—or patient enough to tolerate Fermat's communication style indefinitely. A few words of introduction from Fermat might have kept his less brilliant friends more attuned to the nature and significance of his accomplishments. And he might have found an audience when at last his sense of his discoveries' importance overwhelmed his habitual reluctance to reveal them.

In 1641, Fermat posed this question to Frenicle: being given a number, how many ways can it be the sum of the two small sides of a right triangle? Fermat's question resembled many of the challenges he had sent Frenicle in the past, and by August 1641 Frenicle had been able to find particular solutions to the problem without much difficulty. He even noted that prime numbers of the form $8k + 1$, where k is an integer, are always the sum of the two small sides of a right triangle. (An example is $k = 2$, or $8k + 1 = 17$, which is the sum of 5 and 12, the sides of a right triangle with hypotenuse 13.) Fermat was not terribly impressed with Frenicle's answer, however. He was looking for even broader observations, and he wanted his friend to make the leap he had made.

Fermat's leap was to realize that the algebra in this problem reduces to solving one particular form of the general equation $y^2 - Nx^2 = \pm 1$, where N is a non-square integer. This general form, it turns out, is an equation that has been discovered and forgotten several times over the course of history. Hundreds of years before Fermat, it had fascinated several mathematicians in Asia (notably Brahmagupta in India). Fermat did not know this, however. All he knew was that as he explored triangular and square numbers, answers that reduced to this type of equation kept popping up. By 1643, this equation became the focus of Fermat's research as he searched for a general solution.

Since his discovery, what is now called the Pell Equation has become one of the cornerstones of research in modern number theory, especially because of its ability to reveal the properties of polygonal numbers.

In his own fashion, Fermat was urging Frenicle to move beyond specific solutions to problems like the one involving how many ways can a number be the sum of the two small sides of a right triangle. But Fermat did not spell out his deeper motivation clearly. Because of that, the correspondence of problems between Fermat and Frenicle proved too exasperating for Frenicle to tolerate. In late 1643, he broke off direct communication with Fermat altogether. He never stopped paying attention to Fermat's work, and many years later he still sought news of the problems Fermat posed through other correspondents. But as it turned out, a long time would pass before Fermat had the leisure and focus to communicate regularly with Paris again.

Square Numbers, Triangular Numbers, and the Pell Equation

One way to illustrate Fermat's interest in triangular and square numbers is with billiard balls. Imagine we have an infinite supply of billiard balls, and we want to play variations of pool with different numbers of balls. Usually the game is played with 15 balls arranged as follows:

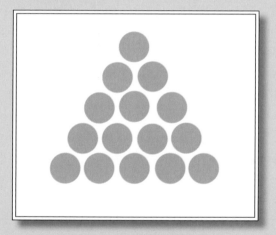

Fifteen ball triangle

Now, if we always want to start with a triangular rack of some size, how many balls can we play with? Here are some examples:

 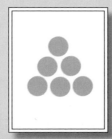 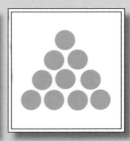

Three, six, and ten ball triangles

The numbers above are all "triangular." In fact, any triangular number could be expressed as a quantity of billiard balls racked in a triangle.

Now, if we also wanted to rack billiard balls in squares, the following would work:

  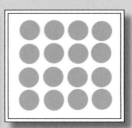

Four, nine, and sixteen ball squares

The question Fermat asks is, essentially, what number of balls could fit into a triangular rack as

well as a square rack? The standard 15 balls will
not work:

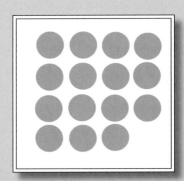

Fifteen ball
square

It turns out that the smallest number of balls
that would fit into both a square and a triangular
rack is 36:

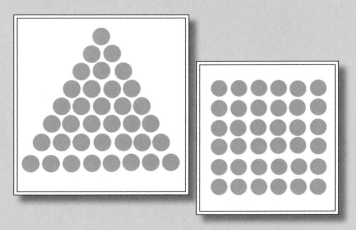

Triangle and square both have thirty-six balls

Fermat, however, was interested not so much in finding answers to individual problems in number theory, but rather in discovering rules that would enable him to solve entire classes of problems. With billiard balls racked in a triangle, we always start with 1 ball on top, 2 in the next row, 3 in the next, and so on. So the number of balls is always something like

1+2 = 3 or 1+2+3 = 6 or 1+2+3+4 = 10 or, say, 1+2+...+8 = 36

Generally speaking, we have

1+2+3+4+...+n

balls, where n is a whole number. To state this generally a triangle with n balls on a side, or the "triangle" of n, is

$$1+2+3+4+...+n = n(n+1)/2 = (n^2 +n)/2$$

To convince ourselves, we might try an example or two:

$$1+2+3+4 = 4(4+1)/2 = 10$$
$$1+2+3+4+...+8 = 8(8+1)/2 = 36$$

A square is always of the form k^2 where k is a whole number. The question is, when do we have $(n^2+n)/2 = k^2$ where both n and k are whole numbers? We saw that 36 works: $36 = (8^2 +8)/2 = 6^2$, so $n = 8$ and $k = 6$.

But how do we find all the possible numbers of balls to play with in either a triangular or a square rack? That question turns out to be significantly harder, but we can at least get a few more particular answers by graphing the solution. Multiplying through by 2, $(n^2 + n)/2 = k^2$ becomes $n^2 + n = 2k^2$. With a bit of insight and algebra, we can rearrange this to be $(2n+1)^2 = 8k^2+1$.

If "$2n+1$" is instead called "x," and "k" is called "y," the equation becomes

$$x^2 = 8y^2 + 1 \text{ or, equivalently, } x^2 - 8y^2 = 1$$

The graph of the function is a hyperbola. The problem of finding numbers that are both square and triangular boils down to finding solutions to the above equation. Once we find out the value of y in the solution, we easily find the value of x (which is $2n+1$) and then solve for n, which is the number of rows in our billiard ball triangle.

We want to see which points on the hyperbola have whole-number coordinates to find solutions to the equation the hyperbola represents (after all, we cannot have a fractional number of billiard balls). So all we do is find the places in which the hyperbola hits a "corner" on our grid. Two of these points are marked. Once we find these points, all we need is the y-coordinate of each of them (notice that point B has y-coordinate 6), and that gives us the value for k. To determine the corresponding value of n, first find the point's x-coordinate, which is equal to 2n+1. At point B this value is 17, which gives us a value of 8 for n.

It turns out that the $x^2 = 8y^2 + 1$ is a particular case of the "Pell Equation," and this sort of problem, which led to knowledge of relationships between triangles and squares, often reduces to one form of the Pell Equation or another. As Fermat gradually came to this realization, he also realized that solving these equations was critical to his quest to probe numbers of different shapes, and this is the main reason they came up so often in his later research.

And the general solution Fermat wanted? He claimed to have found one, but he never revealed it, even when he posed it as a challenge problem in 1657.

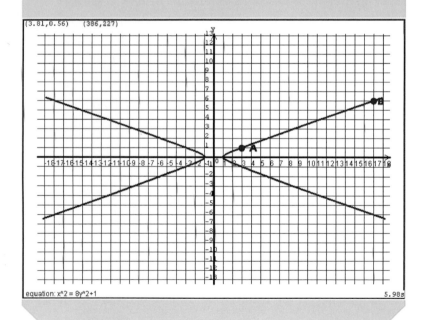

(3.81,0.56)   (386,227)

equation: x^2 = 8y^2+1   5.98s

eight

Last Theorem

For about a decade beginning in 1644, Fermat all but ceased correspondence with his colleagues in Paris. It is not known precisely why. However, historians do know that troubles occupied Toulouse and its *parlement* during this period. There is also some evidence that Fermat's legal career and health suffered periodically, and while he never stopped thinking about math altogether, these problems likely discouraged Fermat from spending time writing to others about it.

By the early 1640s, Fermat had served a dozen years in *parlement* and had been promoted again, to the Chamber of Investigations. His advancement does not seem to have come from any particularly outstanding performance on his part, but more because he had lived through several onslaughts of the plague, a horrific disease that had claimed the lives of many magistrates who were senior to him. Because advancement was largely based on seniority, his survival meant promotion.

Fermat's failures in office may have delayed his advancement at least once, however. He petitioned for another promotion in 1642, but his nomination was delayed. The man through whom he had made the request, Marin Cureau de la Chambre, may have questioned Fermat's abilities before the other magistrates. Fermat mentioned the matter in one of his letters to Mersenne. "I don't know what my standing will be in the mind of M. de la Chambre," he wrote, "since the commission at Castres failed so badly."

He did eventually obtain the desired appointment, and by 1648 he was directed to preside over the higher Chamber of Edicts, a group within *parlement* set up to guarantee justice for religious minorities. But the poor impression he left on Cureau was not an isolated case. Two years before the end of Fermat's life, the first president of *parlement* was under investigation, and Fermat's shortcomings as a jurist came up in a secret report on the president's colleagues. "Fermat, a man of great erudition, has contact with men of learning everywhere," the author of the report, an official of Languedoc province named Claude Bezin, noted. "But he is rather preoccupied; he does not report cases well and is confused." As promotions came, so did his responsibilities. The long hours he once devoted to the world of numbers grew shorter and shorter.

Because time was so precious, Fermat concentrated his efforts on mathematical topics that did not require him to do a great deal of extended writing. Again and again he returned to his edition of Diophantus and its long list of penetrating theorems regarding squares, triangles, and the numbers that could be used to construct their sides. Many of the observations he made in its margins date from these outwardly quiet years, which saw Fermat as inwardly meditative as ever. It may have been during this period that he wrote the brief sentences that came to be known as his Last Theorem.

He spent at least some of his time reading, keeping up on subjects that he had been involved in years before. One of the few letters Fermat sent during the period went to Pierre Gassendi in 1646, responding to the physicist's paper on acceleration. Fermat provided him with a mathematical demonstration of a previous assertion of Galileo's concerning the motion of an accelerating body. But there was no follow-up, no conversation. Fermat had all but disappeared from the scientific community.

He still had friends who had not forgotten him, though. One, a philosopher named Bernard Medon, was among several people who had tried time and again to convince Fermat to publish his work. As always, Fermat refused.

By 1651, Medon was exasperated by this state of affairs. He wrote to their mutual friend Nicholas Heinsius, a prominent Dutch writer, that Fermat's mathematical knowledge was "greater than any mortal possesses," but that "nothing can be extorted" from him. Medon suggested that Heinsius appeal to Queen Christina of Sweden, a famed patron of learning, to request that Fermat share his work.

Heinsius may indeed have tried to enlist the queen's help. If so, the effort did not convince Fermat to abandon his treasured anonymity. Math was his vacation, the world he could visit whenever he needed a break from the everyday. Why take all the pleasure out of it by subjecting it to the same

PETRVS GASSENDVS PREPOSITVS
CATHEDRAL!ᴿ. ECCLESIÆ DINIENSIS

C. Mellan Gall! del et sculp.

Pierre Gassendi *(Courtesy of Mary Evans Picture Library/Alamy)*

kind of judgments and arguments he encountered every day in *parlement*?

Medon also wrote of one of Fermat's closest brushes with death, which came about two years later. Sometime around early 1653 Fermat suffered an attack of the plague, which swept through the countryside every few years and always killed a substantial number of its victims. Fermat's health had been robust up to that point in his life—his letters give only one hint of a slight bout with illness in 1643, when he mentions having had a "discomfort" to Carcavi—but the plague took such a toll on him that it led Medon to report Fermat's death to Heinsius in May 1653.

Fortunately, though, Medon's message had been premature. A short while later, Medon again wrote to Heinsius, saying, "I informed you earlier of the death of Fermat. He is still alive, and we no longer fear for his health, even though we had counted him among the dead a short time ago. The plague no longer rages among us."

Fermat had a family to take care of him during his illness. But his responsibilities toward his family were, by this time, also fairly heavy. He had to think about the education of his five children, and especially of his oldest son, Clément-Samuel, who would eventually inherit his father's position in *parlement*. Fermat had to make sure his eldest son learned how to navigate the difficulties of political life in Toulouse, never an easy task.

There was plenty of dangerous business going on, both inside and outside the walls of *parlement*, and anyone in a position of power could not afford to ignore it. As a rule, magistrates were highly respected. But as representatives of the king's authority, they sometimes had to make decisions

A 1648 battle at the Bastille during the Fronde *(Courtesy of Visual Arts Library (London)/Alamy)*

that were not popular with the masses, and angry mobs were a force to be reckoned with. On one occasion, the scheduled public execution of a prisoner caused riots, and the executioner was killed. To reassert its authority visibly, the entire *parlement* of Toulouse, Fermat included, assembled in full robes and escorted the prisoner through the streets, protected by armed noblemen. Even those at the pinnacle of society had firsthand experience with the dangers of agitating the populace.

An eighteenth-century French *parlement* meeting *(Courtesy of Visual Arts Library (London)/Alamy)*

As if that were not enough, there was frequent unrest in the country, not least during the Fronde, the civil wars that erupted in France between 1648 and 1653. This was a different sort of civil conflict than England experienced in the seventeenth century, as the aim was not revolution, but the protection of the rights of local authorities to retain some political power. Among the local authorities concerned were the *parlements*, of which Fermat's in Toulouse was one of the most established. Aristocrats who struggled to maintain local authority against Louis XIV's desire for absolute power were thrust into a difficult position, for their own power

among the populace was based in large part on their con-
nection to the monarchy and its claim to the divine right to
rule. Though much of the fighting took place elsewhere in
France, the Fronde made it a difficult time to be a *parlemen-
taire* in Toulouse as well.

These troubles magnified the existing internal strife within
parlement itself. Among the magistrates constant squabbles
erupted over questions of legal procedures, distributions of
cases, and the like, chiefly stemming from the desire of each
member to increase his status. Competition resulted in frequent
feuds, sometimes threatening to destroy order altogether.

To a certain extent Fermat's retiring nature, evidenced
by his behavior in the world of mathematics, may have
insulated him from this political infighting. Yet his survival
in the rough-and-tumble world of *parlement* also attests to
his endurance and ambition. Despite his reserved matter in
personal affairs, he managed to maintain his family's social
status and keep his reputation relatively clean through years
of tiresome legal work and continual political storms.

Given the highly charged political atmosphere full of
intrigue and deception, an environment in which he sometimes had to deliver judgment on whether accused men
would live or die, Fermat had to cope with the inconstancy
of human loyalty and will. It is no wonder that Fermat found
enjoyment and solace in the world of numbers, where absolute truths could be discovered.

When, in 1654, Fermat finally emerged from his protracted
silence, it was again in response to a letter bearing a distinguished name and requesting help on a problem. Again,
Fermat would soon turn the conversation back to the fruits
of his own genius.

Fermat's Last Theorem

The forty-eight brief notes that Fermat wrote in the margins of his edition of Diophantus's *Arithmetica* became one of the most inspiring collections of observations about number theory in history. After Fermat's eldest son, Clément-Samuel, published them along with Diophantus's original text in 1670, generations of mathematicians worked to find proofs for these observations. Fermat himself had left little or no justification—only his own reputation for uncanny insights into the properties of numbers. One by one Fermat's observations were proved correct through the efforts of other mathematicians, mostly in the late eighteenth and nineteenth centuries.

One observation remained, however. Though it was probably not the last one Fermat wrote in his *Arithmetica*, it came to be called his "Last Theorem" because it resisted a proof for more than three centuries. In 1994, the British-born mathematician Andrew John Wiles finally proved the theorem.

Diophantus had a problem that concerned breaking down a square number into two smaller squares, which can be expressed as $x^2 + y^2 = z^2$. This was a similar query to the one Fermat had sent to Frenicle. Fermat knew there were an infinite number of whole-number solutions to the problem, but he pondered whether changing the problem

Andrew John Wiles was able to prove Fermat's Last Theorem in 1994.
(Courtesy of © C. J. Mozzochi, Princeton N.J)

could tell him anything deeper about the nature of numbers. Eventually, he realized that the problem, as stated, had a unique property.

In the margin of his copy of the *Arithmetica*, Fermat wrote: "On the other hand, it is impossible to separate a cube into two cubes, or a biquadrate [fourth power] into two biquadrates, or generally any power except a square into two powers with the same exponent." In other words,

$$x^n + y^n = z^n \text{ has no whole-number solutions } (x, y, z)$$
$$\text{for } n > 2.$$

"I have discovered a truly marvelous proof of this, which, however, the margin is not large enough to contain," Fermat wrote in his *Arithmetica*. He apparently never put his proof to paper.

Because Wiles's proof of the Last Theorem ran 150 pages and required sophisticated mathematical techniques that were unavailable during Fermat's time, many people have speculated that Fermat was in error when he wrote his provocative comment about possessing a proof. On rare occasions he did make mistakes—as when he claimed that numbers of the form $2^x + 1$, $x = 2^n$, where n is an integer, are all primes. The eighteenth-century Swiss mathematician Leonhard Euler proved that the result is not prime for $n = 5$, or $2^{32}-1$.

As the years passed, Fermat himself may have sensed that he had not thought through the problem carefully enough before scribbling his note. In a challenge he issued to the English mathematician John Wallis in the late 1650s, Fermat only requested a demonstration that the theorem was true for $n = 3$, not for all possible values of n, which he might have requested had he been in his customary self-assured mood. Fermat's papers show that he possessed proofs for $n = 3$ and $n = 4$, but for no higher values.

Did Fermat possess the "marvelous proof" of which he spoke? The answer will probably never be known for certain, because Fermat left no irrefutable evidence behind. Still, the long search for a proof only increased the mystique surrounding this retiring genius, whose greatest accomplishments went largely unrecognized during his lifetime.

nine

The World Turns Away

When a letter from Paris arrived for Fermat in the summer of 1654, he immediately recognized the sender's last name: Pascal. His old friend Etienne had passed away three years before; now his son, Blaise, was making a name for himself. Blaise Pascal's writings would eventually earn him great fame as a philosopher, but that day he had a question about mathematics for Fermat. It was an unusual one: it concerned the study of gambling. No one had ever successfully examined it with math before.

Pascal did not share his own approach to the problem, but merely said he had a solution—the same sort of "challenge" Fermat himself had often posed to other mathematicians. Fermat had little interest in the topic itself, but Pascal's letter had the effect of drawing him out of a decade of nearly complete silence.

Blaise Pascal *(Courtesy of Visual Arts Library (London)/Alamy)*

Fermat and Pascal's exchange on gambling problems lasted less than three months, but was still enough for them to lay the foundation for the modern study of probability. Though Fermat's interest in probability died quickly, his interest in correspondence had been reborn.

Fermat and the Theory of Probability

During Fermat's time, games of chance were fashionable among the French nobility. Antoine Gombaud, the Chevalier de Méré, had been involved in one such dice game. The exact circumstances of the game are not known, but apparently another gambler had bet that, given a certain number of throws of a single die, he could roll a 6. The other gambler had made several unsuccessful throws when he was called away on urgent business. The question arose as to how to divide the stakes, with the game not yet completed.

Gombaud asked his friend Blaise Pascal to calculate a fair way to split up the money. But he disagreed with Pascal's answer.

When Pascal asked Fermat's opinion, Fermat said that what was needed was to determine all possible outcomes of the game, then define mathematically the likelihood that each outcome would occur. An important principle was that each throw of the die had the same probability of success, regardless of what had happened in previous throws. Each time, the chances of rolling a 6 are thus 1 in 6.

Suppose, Fermat wrote, that "after the stakes have been set, we agree that I do not make the first throw; then, by my first principle, I should receive 1/6 of the stakes as fair compensation." After settling this point, there remains 5/6 of the original stake on the table. If again he did not roll the die, Fermat said, he would be entitled to 1/6 of the remaining 5/6, or 5/36 of the original stakes. The chance of success, then, was the ratio of favorable outcomes to all possible outcomes. Through variations on this principle, which employs the concept of arithmetic series, gamblers could calculate the stakes in different games.

Pascal was relieved that Fermat's analysis agreed with his own. "You have described the play of dice and games with perfect justice," he replied. "I am quite satisfied, and no longer suspect that I was wrong."

The two men exchanged a few more letters on the subject but ran into difficulty when attempting to describe games involving more than two players, and Fermat soon lost interest. Yet the elements of their exchange, however brief, form the basis of classical probability.

Just as he had in 1636, Fermat began to hint at what had been occupying his mind for the previous ten years.

"I hope to send you . . . a sketch of everything worthwhile that I have discovered concerning numbers," he wrote Pascal in late August, mentioning several of his accomplishments. One of these was a proof of his observation that every number is either a triangular number or a number composed of two or three triangles, a square number or one composed of four or fewer squares, a pentagonal number or one composed of five or fewer pentagonals, and so on. He had made this observation next to problem Twenty-nine in Book 4 of the *Arithmetica*, where he also said the margins were too narrow to contain the proof.

In fact, Fermat even hinted there that it might at last be time to begin sharing his work with the world at large. "[This] demonstration, however, which is derived from many varied and most abstruse mysteries of the numbers, cannot be placed here," he wrote in the margin. "For we have decided to direct a treatise and a whole book to this task and to further arithmetic in this direction beyond the Ancients and the known limits in a marvelous way."

Some of the marginal notes in his *Arithmetica* evidently concern the thinking he had been doing along these lines during his decade of near silence. Examining the equation $y^2 - Nx^2 = \pm 1$ had allowed him to see, among other things, that there was a relationship between the value of N and its solubility (its ability to be dissolved). He wrote Pascal that $5x^2 - 1 = y^2$ had a solution because 5 was the sum of two squares, and this fact allowed the easy discovery of a right triangle whose area, added to a given number, equals a square.

Fermat was in as effusive a mood as he had been when he first wrote to Mersenne two decades earlier, but history was not to repeat itself. Pascal, a talented mathematician in his own right, nonetheless failed to display any interest in the subject. "Look elsewhere for someone to follow you in your numerical researches, of which you have done me the honor of sending the statements," Pascal replied in late October. "As far as I am concerned, I confess to you that they go right past me; I am capable only of admiring them and of begging you very humbly to take the first opportunity to [publish them]."

With the wind so abruptly taken out of his sails, Fermat never did put together either a treatise or a book on his new arithmetic. Pascal had nearly lost interest in mathematics altogether by that point in his life, but Fermat, now freshly inspired, had no one else to talk to. He had never really desired publication before, and with Pascal's letter, whatever urge he may have had to do so left him. Two more years would pass before he would again find other mathematicians with whom to correspond.

In 1656, the English scientist Kenelm Digby made a special trip to see Fermat. He brought along a copy of *The Arithmetic of Infinites*, by

English scientist Kenelm Digby visited Fermat in 1656.

After examining English mathematician John Wallis's methods of finding the area under a curve, Fermat composed his *Treatise on Quadrature*, which explained his own techniques more clearly.

John Wallis, who became the best English mathematician of his age. The subject was one Fermat had seen before in his own notebooks: that of finding the area under a curve.

Upon examination of the contents, Fermat must have realized that there was nothing to worry about as far as whose techniques were better; Wallis had not gone nearly as far with his methods as Fermat had many years before. But perhaps Fermat had learned a bit from the dispute with Descartes. Soon after Digby departed, Fermat composed his *Treatise on Quadrature*, which set down his own techniques more clearly than they had been, just as he had summarized his work with curves in his *Introduction to Plane and Solid Loci.* (True to form, though, Fermat simply left the finished treatise among his papers. No one saw it until his son Clément-Samuel published the work in 1679.)

Fermat sensed promise in Wallis's work, though, and hoped the young Englishman would appreciate his work in number theory. In early January 1657, Fermat composed a letter containing two challenge problems: "find a cube which, when added to all its aliquot parts, makes a square" and find "a square number which, added to all its aliquot parts, makes a cube number." He wished the letter to be sent across Europe but also addressed a copy to Wallis personally. Fermat added a hint of nationalistic rivalry to his problems. "We await these solutions which, if England . . . cannot give them, [France] will give," he wrote.

More than a month went by and Fermat received no replies, from England or anywhere else. His problems had failed to inspire anyone, it seemed. In February, he apparently realized he would need to be a bit more explicit than he had been about the reasons he was interested in such problems, for he sent out another challenge problem regarding square numbers, this time with a few sentences of introduction. "There is hardly anyone . . . who understands purely arithmetical problems," Fermat observed. "Is that not because arithmetic has been treated more geometrically than arithmetically? Surely many volumes of both ancient and more recent writers indicate this; even Diophantus indicates it." Fermat proposed to "let arithmetic redeem the doctrine of whole numbers as a patrimony all its own."

Fermat hoped that Wallis would see what others, such as Frenicle, had missed all along. Geometry was a worthwhile pursuit for mathematicians, but it should not so dominate the field that other approaches would be neglected. Geometry had a noble ancestry extending back to Euclid; Fermat was trying to convince his contemporaries that number theory had one too.

His urge to provide an explanation was a wise one, but it was too late to save his first letter, which Wallis had not even received yet. It had been passed from one hand to another via Paris and London before arriving in Oxford, where Wallis was beginning his career as a professor. When it finally arrived in early March, Wallis looked at it briefly and wrote to Lord Brouncker, his English colleague, that Fermat's two challenge problems reminded him of other questions he had come across concerning perfect numbers. And the problems did not interest him. "Whatever the details of the matter," Wallis wrote, "it finds me too absorbed by numerous occupations for me to be able to devote my attention to it immediately. But I can make at the moment this response: the number 1 in and of itself satisfies both demands."

Wallis's answer was trivial—that is, it assumed the simplest value for the variables in Fermat's problems. Wallis had been too busy to see what Fermat was driving at, but Brouncker tried his hand at the problems. However, he came up with answers that involved fractional solutions rather than whole numbers as Fermat had intended. Through Digby a few other responses trickled in as well, including one from Frenicle, who remained interested in Fermat's work even if he was unwilling to communicate directly.

The English responses disappointed Fermat, who wrote back to Digby in late August that he thought Brouncker had not even understood the problem. Though Fermat tried to sound sympathetic, his letter taunted the English mathematic community, and recalled his feud with Descartes. "It is not that I mean . . . to renew the jousts and ancient tilting of lances which the English once carried out against the French," he declared. "Rather, . . . I venture to maintain . . . that accident and luck often intrude into scientific battles as much as in others, and that in any case we can say 'no field can bear every crop.'"

With these echoes of his old conflict with Descartes, Fermat hinted that the English were not capable of pursuing mathematics on his level. Along with this message, he included a critique of Wallis's *Arithmetic of Infinites.* Needless to say, this condescending answer did not endear him to the English mathematical community. In fact, it stirred up another angry exchange between Paris and London that lasted a year. Frenicle and Digby demanded whole number solutions for the original two problems from Brouncker and Wallis, who was himself engaged in a debate with Frenicle over whether a single unit could be considered a cube, a square, or even a number in the first place.

Fermat was for the most part not engaged in this fruitless argument, which went on without him after he had stirred it up. Wallis never grasped why a man of Fermat's talent and international reputation would waste time on such wearying calculations. The point of studying number theory was lost on him.

About the only positive result from the exchange was the joint effort, by Wallis and Brouncker, to find a complete solution to the equation $y^2 - Nx^2 = \pm 1$. Fermat had challenged them to find it at one point, and Wallis included the English pair's result in a volume of his letters that appeared in 1658. Fermat never revealed his solution, though he did mention how he went about finding it the following year, during the course of the last significant correspondence on number theory he would ever have.

Around this time, Fermat tried once more to find an audience for his work with pure numbers. This time, it was with the young Christiaan Huygens, who had been communicating intermittently with Fermat since mid-1656, after Carcavi introduced them by mail.

Christiaan Huygens

At first, Huygens was enthusiastic about exchanging letters with Fermat, whose work with probability had inspired his own paper on the subject. But as time passed, Huygens began to receive letters from Toulouse containing observations about the properties of numbers, just as Pascal had. Huygens tried to be polite to the elderly mathematician. But, as he told Carcavi, he was surprised that Fermat took pleasure in "finding new curves which otherwise have no properties worthy of consideration."

Huygens, who became one of history's great scientists, was interested in mathematics that had immediate application in scientific research, which by this point had developed great momentum across Europe. Fermat's math did not seem to apply to anything practical. Moreover, the letters showed Huygens that the old man had lost touch with current happenings in mathematics. Fermat had become quite isolated.

Sensing Huygens's hesitation, Fermat tried his best to overcome his lifelong habit of secrecy regarding his methods. He composed a full paper, the *Relation of New Discoveries in the Science of Numbers*, in which he not only shared some of his observations on number theory but also described how he had proved these observations. He described how, through a method he called infinite descent, he was able to prove that no right triangle possessed a square area:

> If there were any integral right triangle that had an area equal to a square, there would be another triangle less than that one which would have the same property. If there were a second less than the first which had the same property, there would by similar reasoning be a third less than the second which would have the same property, and then a fourth, a fifth, etc., descending ad infinitum. Now it is the case that, given a number, there are not infinitely many numbers less than that one in descending order (I mean always to speak of integers). Whence one concludes that it is therefore impossible that there be any right triangle of which the area is a square.

He sent the *Relation* to Huygens through Carcavi in 1659, but even then his old habits had not entirely left him. Having outlined the technique, he did not disclose the exact process by which he proved it. Other mathematicians have since done so by their own methods, and it has become one of their most powerful tools for proving difficult ideas.

Fermat even claimed to have proven with infinite descent that his equation $y^2 - Nx^2 = \pm 1$ had an infinite number of solution pairs (x, y) for non-square N. But Huygens, like so many others, could not see what Fermat was driving at. To his eyes the beauty of abstract numbers was no match for the power of practical science. Despite a few rebukes from other senior mathematicians in Paris for neglecting Fermat, Huygens let the correspondence trail off.

Many years would pass before anyone would appreciate what Fermat had done with number theory. Decades later, Leonhard Euler found the equation $y^2 - Nx^2 = \pm 1$ in a book belonging to the English mathematician John Pell. Although

Leonhard Euler mistakenly began to refer to the equation y2 − Nx2 = ± 1 as Pell's Equation after finding it in a book belonging to the English mathematician John Pell.

Pell had done almost no work of his own with the equation, Euler—who at the time was unaware of Fermat's efforts— began referring to it as Pell's Equation. The name stuck. Though it was one of the most penetrating tools Fermat had found for revealing the properties of numbers, by an irony of history it was named for a mathematician who had merely copied it from one of Fermat's letters.

Infinite Descent

One of the most powerful tools that Fermat left for posterity was a way to prove a statement false by showing the impossibility of its ever being true. With a statement that would otherwise require testing an infinite number of potential answers, Fermat realized that he could show the statement's falsity merely by demonstrating that all the potential answers were connected and then proving one of them to be wrong. The technique Fermat invented for doing this, called *infinite descent*, has helped modern mathematicians prove a long list of otherwise difficult theorems.

Infinite descent can be used to prove one of the most jarring discoveries of the Pythagorean mathematicians: that some numbers cannot be expressed as the ratio of two whole numbers. We call these numbers irrational, and ancient Greek history tells us that the first to be found was the square root of 2, or $\sqrt{2}$. We can demonstrate that it is irrational using what we know about right triangles, one of Fermat's favorite objects of contemplation.

Assume first that what we wish to prove false is actually true, namely that $\sqrt{2}$ does, in fact, exist as a ratio of two whole numbers, which we can call a and b:

$\sqrt{2} = a/b$,   or by squaring both sides,
$2 = a^2 / b^2$.   So
$2b^2 = a^2$.

We know that both sides of the equation, whatever specific values a and b turn out to have, represent even numbers. On the left side, we know that multiplying any number, even or odd, by 2 yields another even number. So $2b^2$ must be even. On the right side, we also know that squaring any even number yields another even number, but squaring an odd one yields another odd one (for example, 2 x 2 = 4 but 3 x 3 = 9). Since left and right sides are equal, a^2 must be even, and therefore a must be even as well.

So a, being even, must be twice some other number, c. Plug that into the previous equation:

$$2b^2 = a^2 = (2c)^2 = 4c^2$$

Dividing by 2, we are left with $b^2 = 2c^2$. This shows that b is also even. Let's make the same deduction about b as above: $b = 2d$, so $2c^2 = 4d^2$ or $c^2 = 2d^2$.

We can use this line of reasoning again: c must

be even, so $c = 2e$; and d must be even, so $d = 2f$. From this we can get another pair of equations, $2e^2 = d^2$ and $2f^2 = e^2$. The process can continue as long as we like, unendingly. Since each of these new variables multiplied by 2 equals the previous one, each new one must be smaller (by half) than the previous one. The implication is that, if the square root of 2 is rational, there must always be a smaller pair of whole numbers than the previous pair.

This chain cannot continue indefinitely, of course, because eventually we get to whole numbers that would have to be less than 1, which do not exist. By this logical contradiction, we know that the square root of 2 is not rational.

Fermat described infinite descent to Huygens, but nowhere did he demonstrate it mathematically. He claimed to have proven some of his observations on Diophantus with it, and he may have thought infinite descent could prove his Last Theorem, though—as is so often the case with Fermat—we cannot be sure.

ten
Fermat's Last Days

As Fermat's conversations with Huygens were nearing their end in 1658, a name from his past reappeared. He received a letter regarding his correspondence with René Descartes two decades earlier. Descartes had passed away a famous man in 1650, and Claude Clerselier was compiling his correspondence for publication.

Clerselier had found only two of Fermat's letters among Descartes' papers, and he wrote to Fermat through Digby asking for copies of any other letters that had passed between them during their dispute of 1637-38. Fermat knew he had written no others.

By this point, Fermat was just about at his wit's end as a mathematician. His years of work with analytical geometry had earned him a reputation for brilliance, but others had taken credit for many of his discoveries. His long meditations on number theory had failed to inspire others with either their beauty or their worth. He had patiently spent a lifetime working to illuminate the landscape of numbers, and in his old age he could see it much more clearly and travel it at will. But once there, he found himself alone.

In his frustration, Fermat suspected Clerselier's motives, perhaps wondering if this was yet another Cartesian attack. Jumping to conclusions, he wrote back restating his earlier objections to Descartes' *Dioptrics* and supplying additional criticisms as well. He even made a point of mentioning that Carcavi could supply copies of his complete methods of finding maxima and minima. Fermat was clearly on the defensive.

Clerselier, though indeed a follower of Descartes, had no ulterior motives. Fermat's answers made him jump to a conclusion of his own, however. When Clerselier compared the new responses to the original letters, he thought Fermat was trying to reopen the dialogue about reflection and refraction. The mutual misunderstanding brought on a four-year correspondence between the two men on a subject that had not originally interested Fermat.

Whereas Fermat had been an impartial commentator on the *Dioptrics* twenty years before, he was now venting his long-withheld irritation with Descartes, the man who had effectively undermined his reputation as a thinker. Fermat remained essentially uninterested in the subject of optics. But the exchange led him to think about it again more deeply, allowing him to make one of his greatest contributions to science outside the realm of mathematics.

Descartes' work concerned the way a light beam bends as it travels through different materials. When a ray of sunlight strikes the surface of a pool of water after traveling through the air, the beam clearly bends away from its original path as it passes into the water. The *Dioptrics* explained this behavior with the notion that light somehow travels with different "force" as it passes through different media. Fermat was unwilling to accept this explanation. He was even less willing to accept that Descartes had provided a good mathematical derivation of it, and Fermat's initial criticisms were part of what had touched off the debate so many years earlier.

By 1658, though, Fermat had seen a theory of light that for the most part he did accept. It was from one of his long-time colleagues in *parlement*, M. Cureau de la Chambre. In 1657, Cureau had shown his paper on light to Fermat, who had admired its content. He even suggested to Cureau that together they might prove Descartes incorrect.

"You and I are largely of the same mind," he wrote to Cureau in August, "and . . . if you will permit me to link a little of my mathematics to your physics, we will achieve by our common effort a work that will immediately put M. Descartes and all his friends on the defensive."

Cureau's theory stated in essence that "nature always acts along the shortest path"—or, in the case of light, that it takes the most direct route possible from one point to another, even if that route is not a straight line. Armed with his law of maxima and minima, Fermat set out to do something with mathematics he had long maintained was inappropriate: he would use it to describe the physical world. Cureau's theory would be demonstrated with mathematical rigor, and this would settle the issue with the Cartesians once and for all.

However, Fermat would have to get around a sticky issue before he could do so. He may have agreed with Cureau's theory, but both Cureau and Descartes believed that light travels instantaneously. Fermat disagreed with this idea in principle. When he looked at another aspect of Descartes' arguments that he contested—that light somehow carries a stronger "force" when traveling through a denser material— Fermat realized that he would have to find a better analogy for what was happening.

The inspiration that allowed Fermat to get around these issues (while not alienating his friend Cureau) was his idea that it is the *resistance* of the medium that is the critical factor. Denser materials resist light more effectively, slowing it down and thus determining where the shortest path lies.

To Fermat's great surprise, the mathematical derivation he obtained demonstrated the exact result that Descartes had found years before—what we today call the sine law, or Snell's Law. But here, at last, was a mathematical description of it that would allow the whole issue to be put to rest. Descartes happened upon it, Cureau understood it, and now Fermat had accurately explained it. Clerselier would at last grasp why Descartes' argument was not to the point, and why Fermat had to call attention to his errors. Fermat sent his results to Clerselier on January 1, 1662, confident that the old debate would end at last.

Fermat's efforts are today seen as the first accurate explanation of how light passes through different media, and they have influenced subsequent work in optics. But Clerselier was unwilling to see things Fermat's way. He shot back that the idea that nature acts along the shortest paths was not a physical principle but merely a moral judgment about the nature of the physical world. Clerselier was incorrect, and Fermat knew it. But the squabble sputtered on until the end of Fermat's life.

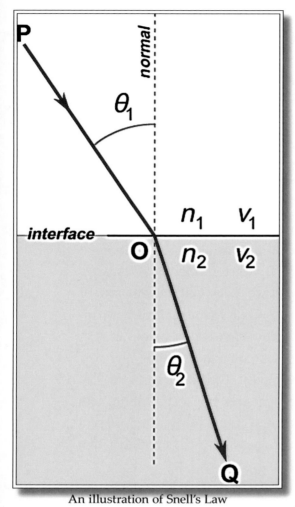

An illustration of Snell's Law

By 1660, Fermat was an old man by the standards of his day. His bout with the plague several years before and the stresses of daily life, not to mention the frustrations his mathematical correspondence had brought, all began to catch up with him. Ill again that year, he drew up his will in March, naming son Clément-Samuel his sole heir. This does not necessarily imply that Fermat had disputes with his other children, it was simply that leaving all property to one's firstborn son was common practice for the day. (His other son, Jean, went on to become an archdeacon, and his daughter Claire eventually married; his two youngest daughters, Catherine and Louise, both born when Fermat was in his forties, became nuns.) In December of 1661, Fermat began the process of designating Clément-Samuel as his official successor in *parlement*.

Fermat continued to serve as a magistrate and write an occasional letter, though by 1662 he had written the last

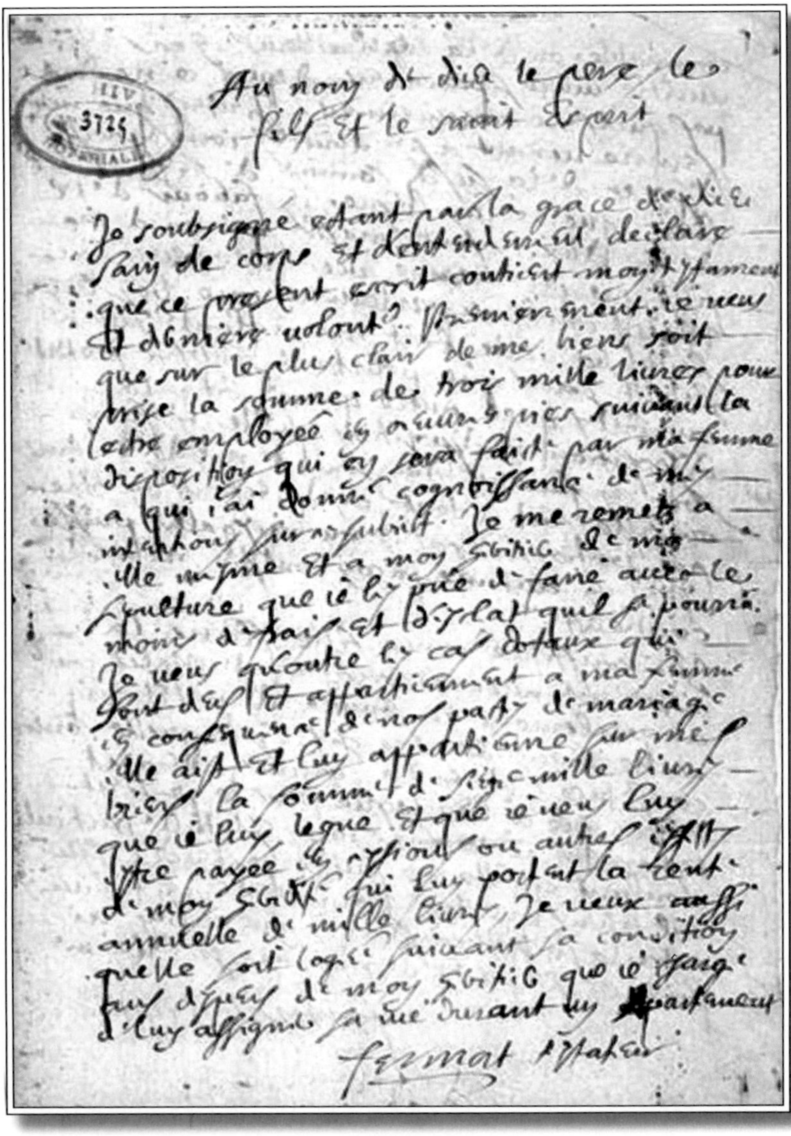

Fermat's handwritten will

correspondence that revealed any new information about his thoughts. He remained in contact with Clerselier until 1664, still engaged in their dispute over optics. Despite his lifelong unwillingness to engage in such arguments—especially when they involved subjects about which he admitted to having no interest or expertise—Fermat seemed unable to avoid them, right up until the end.

The end came during the winter of 1665. On January 9, he signed his last piece of legal paperwork. He died three days later, at sixty-three years old, and was buried in the Church of St. Dominique in Castres. His wife survived him by more than twenty-five years, but even she would pass away long before the mathematical world paid much attention to the work Fermat had left behind.

Pierre de Fermat was not a man who intended to change the world. His efforts, from his earliest days in Bordeaux, were focused on reconstructing knowledge that had been lost, not with breaking new ground. But break new ground he did. In attempting to rebuild the discipline of geometry, Fermat created a system to rebuild it using algebra, uniting the two fields in a manner still central to our thinking today. In attempting to answer a question Pascal posed, Fermat took the first steps toward understanding probability. And in a lifelong attempt to understand the deeper nature of the numbers he loved meditating upon, Fermat single-handedly revived the ancient discipline of number theory, a legacy that has inspired mathematicians for centuries. Those who truly understood the appeal of this work would not arrive until many decades after his death.

Many of Fermat's contemporaries, especially Descartes and Pascal, seem highly emotional and even full of bluster in comparison with the magistrate from Toulouse, whose letters invariably display a calm tone. But Fermat's own character had conflicting aspects as well. He sometimes alluded to the community of scientists freely sharing knowledge with one another, an ideal that the English philosopher Francis Bacon had popularized. Yet he repeatedly withheld his methods from other mathematicians, sometimes seeming to taunt them with reference to his own solutions.

During his three decades in Toulouse, Fermat never made time to visit Paris or to write more than a few of the major works he planned. Though he found a few correspondents whose interest inspired him to set down a handful of his best ideas, how he developed his great talent with number theory will probably remain a mystery forever. And Fermat's reluctance to publish meant that most of the contributions he is known to have made entered history under other people's names. Analytic geometry is now called Cartesian. Beaugrand passed off Fermat's method of finding tangents as his own. The technique for finding maxima and minima is now called "De Sluse's Rule." And the relationship that showed Fermat so much about the properties of numbers would be called "Pell's Equation," after a mathematician who did little more than copy it.

But Fermat was well known to his contemporaries, and his letters were copied again and again, passing from one hand to another across Europe. By the year of Fermat's death, his ideas reached one of the two men who would create calculus, the revolutionary form of math we still use to describe movement and change.

Though many of his ideas are important to calculus, Fermat did not invent it, nor is his name well known for his contributions to it. He was a problem-solver who was not looking to describe physical behavior with math—though, as his contribution to optics demonstrates, he could think in physical terms when the need arose. However, had he not been involved in the dispute with Descartes, he probably would never have done so.

The names we justifiably associate with the calculus are Isaac Newton and Gottfried Wilhelm Leibniz, whose followers engaged in a bitter dispute over which of the two had the right to be called its inventor—a dispute that recalls the one

Sir Isaac Newton based his system of calculus on Fermat's method of drawing tangents.

between Fermat and Descartes. But Newton was well aware of whose shoulders he had stood upon to create his system, though it took centuries to find evidence of this. In 1934, Louis Trenchard Moore discovered a note in which Newton, one of the greatest scientific minds in history, wrote that he developed calculus "based on Monsieur Fermat's method of drawing tangents."

Today Fermat is remembered most for his unparalleled understanding of the subject he loved best: the patterns within numbers themselves. His work's very inscrutability inspired many illustrious mathematicians—among them Euler and Gauss—to find their own proofs of Fermat's findings and turn number theory into the legitimate field it is today. Fermat's name made the front page of the *New York Times* in 1994, when Andrew Wiles finally proved Fermat's infamous Last Theorem correct. But the theorem itself is merely the one piece of a vast body of work created by one of the most intuitive thinkers in the history of numbers.

Timeline

1601 Born around August 17 in Beaumont-de-Lomagne, France.

Late 1620s Spends time in Bordeaux, encounters work of François Viète; formulates ideas on analytic geometry, grows interested in number theory; meets Jean Beaugrand and Etienne d'Espagnet.

1631 Returns to Toulouse, takes post in *parlement*; marries Louise de Long.

1636 Begins corresponding with Marin Mersenne; works circulate in Paris.

1637 Sends criticisms of Descartes' *Dioptrics* to Mersenne upon request in the fall.

1638 Promoted to serve in Chamber of Investigations in *parlement*; begins dispute with Descartes concerning method of determining maxima and minima; begins corresponding on number theory problems with Pierre Brûlart.

1640 Corresponds with Bernard Frenicle de Bessy on number theory problems; writes *Analytic Investigation of the Method of Maxima and Minima*.

1642 Promoted to serve in *parlement*'s criminal court.

1643   Writes letter to Brûlart detailing the use of *adequality* to find maxima and minima; end of regular correspondence with Paris group.

Late 1652 or early 1653   Suffers attack of plague and nearly dies.

1654   Promoted to the Grand Chamber, the highest chamber of *parlement*; corresponds with Blaise Pascal on subject of probability.

1656   Visited by Kenelm Digby, who later becomes intermediary in correspondence with British mathematicians; begins correspondence with Christiaan Huygens on probability and number theory.

1657   Issues "challenges" to John Wallis and others on complete solution of the Pell Equation.

1659   Writes to Huygens concerning infinite descent and number theory.

1661   Draws up last will and leaves position in *parlement* to firstborn son, Clément-Samuel.

1665   Dies January 12 in Castres, France.

Sources

CHAPTER THREE: Introduction to the Paris Circle

p. 37, "Now if I wished . . ." Michael S. Mahoney,
The Mathematical Career of Pierre de Fermat (Princeton,
NJ: Princeton University Press, 1994), 288.

p. 39, "If M. d'Espagnet . . ." Ibid., 54.

p. 40, "Let NZM be a straight line . . ." Ibid., 84.

CHAPTER FOUR: A Stream from the South

p. 51, "Let A be the unknown . . ." Ronald Calinger,
ed., *Classics of Mathematics* (Englewood Cliffs, NJ:
Prentice Hall, 1995), 377.

CHAPTER FIVE: A Comedy of Errors

p. 61, "when I consider . . ." Mahoney, *Mathematical
Career*, 172.

p. 63, "His rule . . ." Ibid., 180.

p. 65, "M. des Cartes is right . . ." Ibid., 191.

CHAPTER SIX: The Lure of Pure Numbers

p. 71, "or the next one following," Mahoney, *Mathematical
Career*, 293.

p. 71, "I have several shortcuts . . ." Ibid., 294.

p. 72, "to raise a great structure," Ibid.

p. 73, "From these shortcuts . . ." Ibid.

CHAPTER SEVEN: The Shape of Numbers

p. 79, "I am led to believe . . ." Mahoney, *Mathematical Career*, 308.

p. 84, "One should propose . . ." Ibid., 313-14.

CHAPTER EIGHT: Last Theorem

p. 95, "I don't know what . . ." Mahoney, *Mathematical Career*, 18-19.

p. 95, "Fermat, a man of great . . ." Ibid., 20.

p. 96, "greater than any mortal . . ." Ibid., 24.

p. 98, "I informed you earlier . . ." Simon Singh, *Fermat's Enigma: The Epic Quest to Solve the World's Greatest Mathematical Problem* (New York: Walker and Co., 1997), 37.

p. 103-104, "On the other hand . . ." Amir D. Aczel, *Fermat's Last Theorem: Unlocking the Secret of an Ancient Mathematical Problem* (New York: Delta, 1996), 9.

p. 104, "I have discovered . . ." Ibid.

CHAPTER NINE: The World Turns Away

p. 110, "I hope to send you . . ." Mahoney, *Mathematical Career*, 333.

p. 110, "[This] demonstration, however . . ." Ibid.

p. 111, "Look elsewhere for someone . . ." Ibid., 334.

p. 113, "find a cube which . . ." Ibid., 337.

p. 113, "a square number which . . ." Ibid.

p. 113, "We await these solutions . . ." Ibid.

p. 113, "There is hardly anyone . . ." Ibid., 338.

p. 113, "let arithmetic redeem . . ." Ibid., 339.

p. 114, "Whatever the details . . ." Ibid., 341.

p. 115, "It is not that I . . ." Ibid., 342.

p. 117, "finding new curves . . ." Ibid., 67.

p. 116, "If there were any integral . . ." Ibid., 348.

CHAPTER TEN: Fermat's Last Days

p. 124, "You and I are largely . . ." Mahoney, *Mathematical Career*, 394.

p. 131, "based on Monsieur Fermat's method . . ." Singh, *Fermat's Enigma*, 44.

Bibliography

Aczel, Amir D. *Fermat's Last Theorem: Unlocking the Secret of an Ancient Mathematical Problem.* New York: Delta, 1996.

Beik, William H. "Magistrates and Popular Uprisings in France Before the Fronde: The Case of Toulouse." *Journal of Modern History* 46, no. 4 (December 1974): 585-608.

Bell, E. T. *Men of Mathematics.* New York: Simon and Schuster, 1937.

Berlanstein, Lenard R. *The Barristers of Toulouse in the Eighteenth Century (1740-1793).* Baltimore: Johns Hopkins University Press, 1975.

Calinger, Ronald, ed. *Classics of Mathematics.* Englewood Cliffs, NJ: Prentice Hall, 1995.

Devlin, Keith. *Mathematics: The New Golden Age.* New York: Columbia University Press, 1999.

———. *Mathematics: The Science of Patterns.* New York: Scientific American Library, 1994.

Gittleman, Arthur. *History of Mathematics.* Columbus, OH: Charles E. Merrill, 1975.

Heath, (Sir) Thomas L. *Diophantus of Alexandria: A Study in the History of Greek Algebra.* New York: Dover Publications, 1964.

Katz, Victor J. *A History of Mathematics: An Introduction.* 2nd ed. Reading, MA: Addison-Wesley, 1998.

Kline, Morris. *Mathematical Thought from Ancient to Modern Times*. New York: Oxford University Press, 1972.

Mahoney, Michael S. *The Mathematical Career of Pierre de Fermat*. Princeton, NJ: Princeton University Press, 1994.

Peterson, Ivars. *The Mathematical Tourist*. New York: W. H. Freeman and Co., 1988.

Schneider, Robert A. *The Ceremonial City: Toulouse Observed 1738-1780*. Princeton, NJ: Princeton University Press, 1995.

———. *Public Life in Toulouse 1463-1789: From Municipal Republic to Cosmopolitan City*. Ithaca, NY: Cornell University Press, 1989.

Singh, Simon. *Fermat's Enigma: The Epic Quest to Solve the World's Greatest Mathematical Problem*. New York: Walker and Co., 1997.

Stewart, Ian. *From Here to Infinity*. Oxford and New York: Oxford University Press, 1996.

Struik, D. J., ed. *A Source Book in Mathematics, 1200-1800*. Princeton, NJ: Princeton University Press, 1986.

Young, Laurence. *Mathematicians and Their Times*. Amsterdam: North-Holland, 1981.

Young, Robyn V., ed. *Notable Mathematicians from Ancient Times to the Present*. Detroit: Gale, 1998.

Web Sites

http://www-history.mcs.st-andrews.ac.uk/Biographies/Fermat.html
A short biography of Pierre de Fermat written by J. J. O'Connor and E. F. Robertson of the School of Mathematics and Statistics, University of St Andrews, Scotland.

http://www.simonsingh.net/What_is_the_Theorem.html
An explanation of Fermat's Last Theorem from Simon Singh, author of *Fermat's Enigma: The Epic Quest to Solve the World's Greatest Mathematical Problem.*

http://www.pbs.org/wgbh/nova/proof/wiles.html
The PBS series *NOVA* interviewed Andrew Wiles, who proved Fermat's Last Theorem in 1994.

Glossary

Aliquot parts
The proper divisors, or factors, of a number.

Analytic geometry
The use of algebraic equations to describe and study curves (as distinguished from classical geometry, which did so primarily by physical construction).

Conic section
A locus that can be created by slicing a cone through a plane at various angles of intersection. These can all be represented by first- or second-degree indeterminate equations in two unknowns, such as the parabola $y = 2x^2 + 3$.

Determinate equations
Equations that have a single solution. For example, $x + 2 = 5$ is determinate. See *indeterminate equations*.

Friendly numbers (also called amicable numbers)
A pair of numbers, the sum of each of whose aliquot parts equals the other number.

Indeterminate equations
Equations that have an infinite set of solutions. For example, y

$= x^2$ is indeterminate. See *determinate equations.*

Locus (pl. *loci*)

A unique curve defined by an equation. Fermat would consider lines, hyperbolas and cones to be *loci.*

Maxima and minima

Points at the extreme values of a curve, such as the vertex of a parabola.

Number theory

The study of the properties of numbers themselves, independent of what those numbers represent. Searching for prime numbers has been a chief occupation of number theorists from Greek times until today.

Perfect number

A number, the sum of whose aliquot parts equals itself; 6 is perfect because it is evenly divisible by 1, 2 and 3.

Restoration

The recreation of the lost texts of history, an effort that concerned many mathematicians of the seventeenth century. Fermat generally "restored" ancient Greek mathematical knowledge by taking problems from classical sources and solved them with modern algebraic methods unknown to the Greeks, often providing general knowledge that could be applied to an entire class of problems.

Tangent

A line is tangent to a curve if they both pass through a single point in the same direction.

Theory of equations

A branch of mathematics that, in Fermat's day, concerned the transformation of equations of many different types into a relatively small number of standard forms that could be solved by straightforward algebraic methods.

Index